The Coral Seas

The Coral Seas

Andrew C. Campbell

G. P. PUTNAM'S SONS
New York

© Orbis Publishing Limited, London, and IGDA, Novara 1976
SBN: 399-11778-4
Library of Congress Catalog Card Number: 76-6036
Printed in Italy by IGDA, Officine Grafiche, Novara

Title page: School of white grunts sheltering among corals (Dick Clarke/Seaphot)

Contents

Life and death of the reef   7

The coral environment   31

Diving in coral seas   55

The fish of the coral reefs   73

Tourism and coral harbours   85

Men of the coral islands   93
by C. R. Hallpike

Protecting the coral heritage   123

Bibliography   128

Index   128

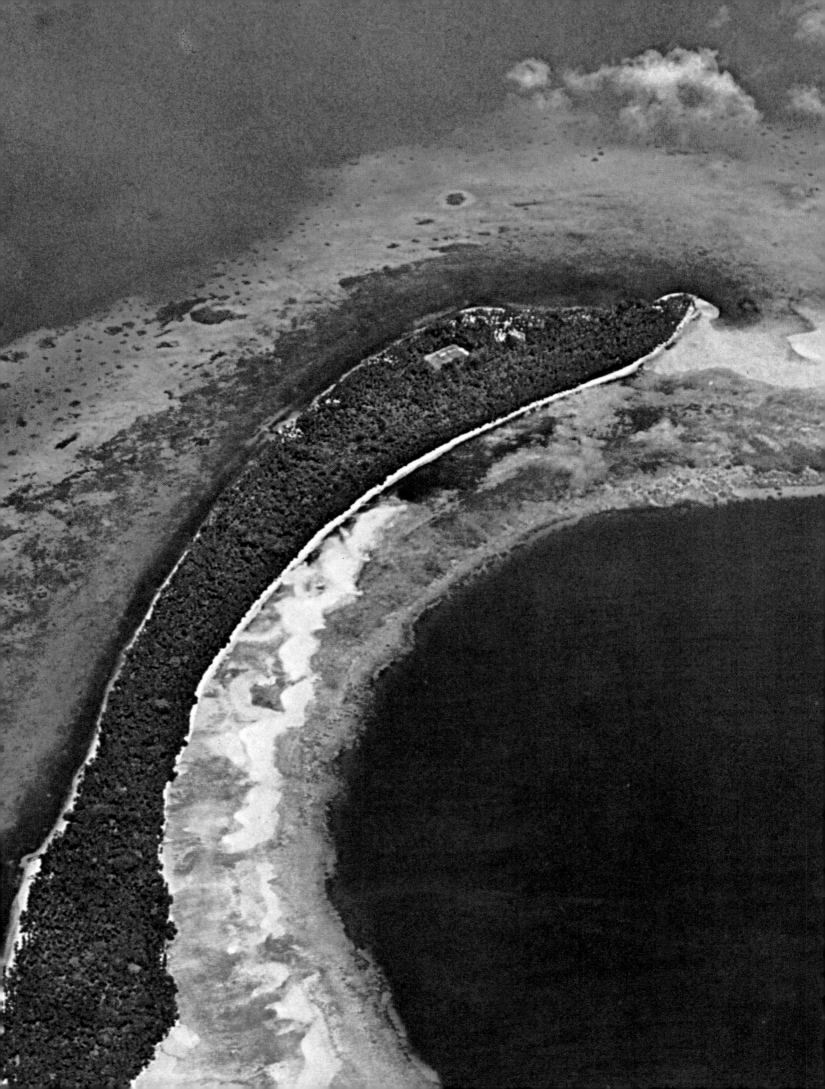

Life and death of the reef

Left: An aerial view of the peninsula of Japen Island, off New Guinea. It is possible to see the coral formations on the sea-bed surrounding the peninsula as well as the dense vegetation on land

Following pages: Waves breaking over the reef crest at a point on the Great Barrier Reef. The change from very deep to very shallow water is indicated by the change of colour from dark blue to paler blue

Most people have had dreams of life on a coral island, but very few would be able to describe one with any accuracy. Popular books and broadcasts have continued to create a romantic image of these islands which, if one sees them from sea level, is not always entirely justified. Once beneath the waves, however, the coral island shows itself as a fantastic and very beautiful world, depending entirely upon a complex web of interrelationships between plants and animals.

Coral islands are so called because they have developed as a direct result of the growth of vast numbers of coral animals. These animals, known as true corals or madreporarian corals, are closely related to the familiar sea-anemones, and occur in all the oceans of the world, both cold and warm. It is only in the tropics, however, where the sea temperature seldom falls below 22°C (72°F), that true corals are able to form reefs. This reef-building is possible because corals, unlike the related sea-anemones, secrete a cup-like skeleton of limestone in which they sit. Under certain circumstances the coral animal, known as a polyp, can completely withdraw into the skeletal cup. When the coral dies its skeleton remains and another organism, possibly a coral, may grow on top of it. Over many thousands of years all the collected coral skeletons, together with those of associated organisms like coralline algae, form an underwater mound or reef.

The factors which govern reef formation are numerous, and scientists are far from agreed in understanding how and why reefs form. An important contributory element is that most reef-building corals depend on a good supply of sunlight for the formation of their skeletons. This means that they cannot grow in deep water where light cannot penetrate or in cloudy conditions. If a suitable firm base is available in clear, well-illuminated waters at the correct temperature, corals will begin to grow and may eventually form a reef.

The earliest theory for the origin of coral reefs and islands was that put forward by the French poet and biologist von Chamisso in about 1818. He considered that coral grew until it reached the sea's surface where it stopped. The ocean would then keep the coral on the outside of the reef alive, but that in the centre would die and be eroded away by the sea so that a lagoon is formed, surrounded by what we know today as an atoll, a ring-shaped or horseshoe-shaped reef with an inner lagoon.

Two French naturalists, Quay and Gaimard suggested in 1825 that reef-building corals flourished in shallow water. They proposed that if such corals grew around the rim of a submerged volcano this would explain atoll formation.

The famous nineteenth-century biologist Charles Darwin made observations on coral reefs when he sailed on his epic voyage of discovery aboard H.M.S. *Beagle* from 1831 to 1836. Darwin believed that coral reefs grew first as fringing reefs in shallow water following the outline of the coast. Subsequent geological upheavals had led to the sinking of a coastline, thus placing these fringing reefs in progressively deeper water. Because corals always grow towards light they would, assuming the subsidence to be a slow and progressive phenomenon, keep pace with it and eventually overtake it, thus reaching the surface from much deeper water. By this explanation Darwin accounted for the formation of off-shore reefs. Darwin's theory received support from

Above: Dense mangrove vegetation on a tiny coral island off the eastern coast of New Britain. The floating mangrove seeds are dispersed by the ocean currents

another famous nineteenth-century student of corals J. D. Dana, but not every scientist agreed.

Sir John Murray, who was leading biologist on H.M.S. *Challenger* during the oceanographic expedition of 1874-1876, proposed that corals grew on the crests of submarine banks or 'hills' where the previous accumulation of mineral or other sediments had brought these within the depth limits of coral growth. Once established, the corals would grow up and form a ring-like reef around the bank. If the submarine bank actually protruded from the sea an island would be there anyway. If not, the gradual accumulation of silts and deposits within the ring of coral would fill it up and form an island. If the inner rock was softer than the reef coral limestone, the natural course of weathering might lead to its erosion and the formation of an atoll.

Various other theories for the origins of coral reefs and islands were proposed by geologists and biologists in the late nineteenth and early twentieth centuries. Like Darwin and Murray, however, the scientists who proposed them were hampered by a lack of suitable equipment and the experience to put them to the test. The Submerged Bank Theory proposes that coral assemblages grow on flat surfaces, or banks, as they become submerged, either due to the sinking of the land or a rise in the sea level. Alternatively the Daly Glacial-Control Theory tries to account for the phenomenon as a result of the way in which the ice caps of the poles absorbed sea-water as they extended outwards during the development of the last ice age a million years ago. This taking up of water to form ice resulted in a lowering of the levels of the oceans and the consequent establishment of new shore lines. These new shores were eroded by the action of the seas and provided platforms which were suitable for the development of coral when the ice melted and the sea levels rose once again.

Modern evidence, which comes from the investigation of drillings, marine sediments, and from the carbon-dating of rocks and fossils, suggests that in many regions of the world where corals are found, the sea-bed was higher in former times, and that it was subsequently sunk. Regardless of which theory is correct, it is almost certain that various processes have contributed in different geographical situations, and that one simple explanation will not account for all the different formations. Nevertheless, the recent evidence for subsidence of the sea-bed does lend support to Darwin's views.

The speed of coral growth is also relevant to these theories. Growth rates vary enormously

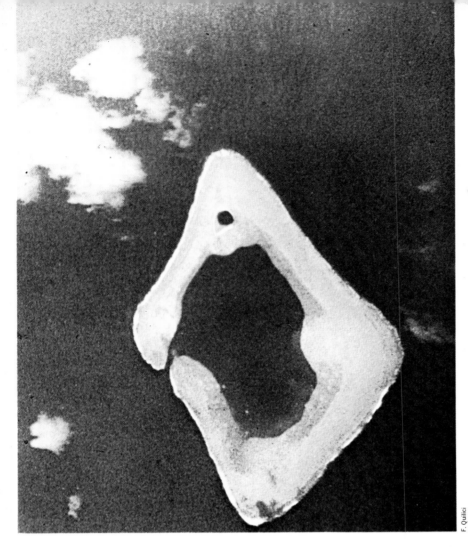

according to the species of coral and its environmental conditions. Estimates have varied from as little as 5 mm (0·2 inches) per year for the more massive forms to as much as 200 mm (8 inches) per year for branches of the bushy varieties. Obviously it is difficult to put an exact figure on the age of reefs, but some authorities have suggested that many reefs now living have been formed within the last 30,000 years. Others believe that in certain cases the time factor may be up to five times greater.

The structure of a typical coral island can best be summarized as a relatively soft area of sand or silt lying within a shallow lagoon which is enclosed by a ring-like reef. From this basic form many varieties have been developed. The outermost edge of the reef is covered by the very thin layer of live coral polyps which are all generating calcium carbonate to sustain their chalky skeletons. As the waves of the sea pound on the growing reef some of the coral limestone is broken off and washed over the top of the reef to be ground to sand and powder by storms. These grains of coral sand constitute the soil of the island which contrasts markedly with the harder rock of the reef itself. As the sand in the middle of the island is periodically exposed to rain and

Above: Aerial view of a Polynesian atoll, showing the distinctive reef enclosing the lagoon which connects with the sea by a narrow channel

Right: Formation of an atoll, after Darwin. A. Volcanic island with small fringing reefs beginning to form. B. As the island sinks, the reefs keep pace and remain at the surface. C. As the process continues, sedimentation builds up the sea-bed around the reefs. D. Final formation of an atoll with lagoon

Following pages, left: Series of coral outcrops forming a perforated reef around an island in the Philippines

Following pages, right: View of Raiatea Island in the Society Islands. Between the outer edge of the reef and the shore is a lagoon

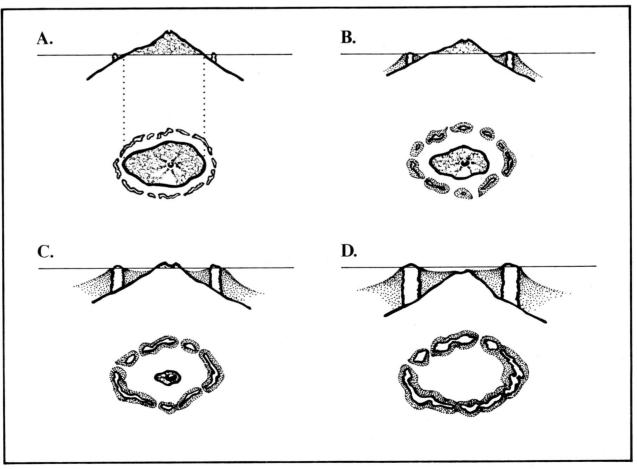

intense heat, certain chemical changes take place within it. Some of its components are then washed down into the deeper parts where, aided by heat and pressure, they re-crystallize to form rocks. In this way a number of physical and chemical processes augment the biological ones in establishing the coral island.

There are islands which appear to be coral islands, yet a closer investigation shows that their rocky cores have all the characteristics of the nearby continental land masses. The most probable explanation is that such an island has been separated from the land mass by changes in water or land levels, and has since developed its own fringing reefs in one of the ways already discussed. The important point, however, is that the reef has formed because of the island's existence, rather than the island's formation being due to the reef. These islands are termed 'continental islands' and show very different characteristics to atolls or 'sand cays'. They are frequently much older and rise a long way out of the sea. The term 'sand cay' is used to describe the low mound of sand which forms on top of some coral reefs and hence constitutes a simple island.

Since true coral polyps live attached to rocks and shells and are unable to move about they must rely on currents of water to bring their supplies of food and oxygen to them. For this reason the site on which a coral polyp grows is very important, and a number of conditions have to be satisfied in order that corals may develop. These include position with regard to currents and tidal movements, the temperature and illumination of the sea, freedom from smothering sediments and freedom from predators.

Some corals are hermaphrodites, that is they have both male and female sex organs within each

Below: Crown-of-thorns starfish, which feeds on live coral colonies and has become notorious in recent years for the damage it causes

G. Gualco

Above: A specimen of Crown-of-thorns lodged in a colony of fire-coral (Millepora) so that its tube feet can be seen on the underside

Above right: Crown-of-thorns starfish moving over a brain coral

polyp, while others exist as males and females. In either case sperms are normally released into the surrounding sea-water and enter another polyp via the mouth to fertilize the eggs. When reproduction takes place in mature polyps, the fertilized eggs develop as embryos inside the adult's body. At a certain stage these embryos are released as larvae which form the dispersal stage of the coral's life-cycle. A coral larva generally consists of a minute cluster of cells known as a planula. The outer surface of the planula bears minute filaments called cilia which are able to beat back and forth, so imparting a swimming movement to the planula. Because of the size of the cilia, the planula's ability to move is very restricted, but this action is sufficient to bring it into contact with currents of water which may sweep it along for considerable distances. While the planula is effectively planktonic, or drifting, it is exposed to a variety of risks, both from predators and from currents which may carry it into unsuitable areas. When the larvae mature their behaviour changes and they start to explore the sea-bed in order to find an appropriate point on which to settle. Only a small proportion of them reach this stage because of the hazards of the planktonic stage. When finding a settlement point, many larvae take into account the surface texture of the substratum, the presence or absence of certain chemicals, and the presence or absence of other types of animals and bacteria. Having selected a suitable point a metamorphosis takes place, and the planula turns into a small, juvenile coral polyp which immediately begins to secrete a calcareous skeleton and to live like the adults.

At any given moment there can only be a limited amount of free space available on suitable rocks or shells for coral larvae. The larvae of various types of coral will therefore often compete for this space when breeding takes place, as will the settling larvae of various other sessile reef animals like sea-mats and sea-squirts. This competition is a very important factor in controlling the way a reef develops, and because certain species of coral can metamorphose and survive better in particular situations on a reef, such as on the outer reef crest, the coral community develops in characteristic zones. This is also true for many of the reef organisms, and a close examination of a reef will show characteristic species in characteristic situations. Many coral reefs grow in tidal seas, and the effect of the tides is to produce a zonation of organisms on those parts of the reef which are both exposed and covered by the tides twice each day. The situation on the top of a reef is in many ways comparable with that which may be encountered on a temperate shore, where the positions that organisms occupy are directly controlled by the degree of exposure to air that they can tolerate.

So long as the other conditions, both physical and biological, necessary for the life of newly metamorphosed coral are satisfied, it will grow on food swept towards it in the currents that surround it. This food will be made up mainly of planktonic organisms, including the developing larvae of a variety of animals.

Most reef-building corals are colonial. This means that a number of polyps develop by budding from the original or ancestral individual, and come to live side by side – each contributing to the bulk of the colony by secreting its own skeleton. When the first polyp which

metamorphosed from the larva has reached a certain size (the size varies according to species), this budding process will begin, and the daughter polyp will, in its turn, produce another when mature enough to do so. In this way thousands of polyps may be built up over a period, and all these polyps will be able to produce sperms and eggs when mature and thus complete the life-cycle by the formation of more embryos and larvae.

The reef formation depends on more than the contribution made by the true corals which have been described so far. A variety of other organisms contribute to the aggregation of limestone by the deposition of their skeletal remains. These remains include mollusc and crustacean shells, sponge spicules, the skeletons of starfish and sea-urchins, worm tubes and other animal material originating from the reef community

Left: Crown-of-thorns starfish can, according to some reports, destroy around 300 square centimetres (46 square inches) of live coral in 24 hours. The folds of the partially everted stomach as well as the tube feet can be seen here

Right: Live polyps of the true coral Goniopora. *The polyps of this coral are able to extend a long way out of the skeleton. There is a clear difference in the appearance of the extended and withdrawn individuals*

Left: Detail of one arm of a Crown-of-thorns showing spines, tube feet and an orange-red eye spot (a simple light receptor)

Right: The skeleton of a sea-fan exposed after the death of the outer layer of living polyps

itself, as well as sediments that may be swept there by the sea. One important factor is the coralline algae. These seaweeds are quite unlike the familiar sea-shore species. They consist of small to moderate-sized knobby and plate-like growths which are quite as hard as the coral skeletons themselves. A typical example is the genus *Lithothamnion* whose members comprise hard chalky nodules coloured pink to grey in life due to pigments present in the cells. Like the corals they extract calcium carbonate from the sea-water which surrounds them and incorporate it into their structures. They may often be encountered on quite exposed parts of the reefs and by their growth patterns they help to bind the corals and other skeletal remains together, thus making a very important contribution to the reef.

Just as a variety of physical and biological

factors contribute to the formation of a coral reef, as great a variety are involved in its breakdown. For a reef to grow and remain healthy the coral polyps which are maintaining its outermost surface of limestone must be able to grow and reproduce. Any shift in environmental conditions like temperature and salinity may seriously affect the polyps' ability to do this at the rate required to keep pace with other agencies which are constantly attacking them.

Coral colonies, especially the branching types like *Acropora*, some types of which are known as elkshorn or stagshorn coral, are particularly vulnerable to the action of heavy waves and to the effects of storms and cyclones. Although delicate branching forms are not able to grow in the surf-zone of a reef they can grow in profusion where the water is less turbulent. Severe storms can still break them up, however, and they either fall into

Above: Specimen of Crown-of-thorns feeding on a branching colony of Acropora. *The whitened skeleton is a clear indication that the starfish has killed and digested the soft polyp tissue leaving only clean calcium carbonate*

Left: The Great triton is a mollusc belonging to the whelk family and an occasional predator of the Crown-of-thorns. Here it is shown in the act of consuming one

deeper water or are washed on to the top of the reef itself where they become further broken and ground into rubble. On certain reefs which are frequently exposed to this sort of weather a characteristic rubble rampart may be found.

Living coral polyps are particularly susceptible to sediments which can sometimes smother them, clogging up the mouths and tentacles and hindering respiration and feeding. In recent years several incidents of the destruction of vast numbers of corals by heavy sedimentation have been reported following dredging activities in harbours and estuaries. Such human intervention on the sea-bed has often caused new currents and patterns of sedimentation, clearly demonstrating the way in which interference at a point which may be entirely remote from a reef can nevertheless do great damage to it.

Fluctuations in water levels can also have a serious effect on the well-being of corals and reefs. In some waters, the central parts of the Red Sea for example, there are virtually no tides. In summer there is an increased rate of evaporation at the surface due to the very high temperatures. This loss of water causes a lowering of the sea level by about 200 mm (8 inches), thus exposing the tops of some of the reefs and killing all the organisms that are exposed. In these regions such changes appear to be a regular hazard to life in shallow water.

Like other organisms, coral polyps fit into food chains or food webs. These terms are used by ecologists to illustrate the interdependence of one species on another for food. Species A, for example, may be herbivorous, feeding on seaweeds, while species B is carnivorous and feeds on species A, and yet other species of higher carnivores will feed on B. Generally, there are fewer

Below: The Great triton is a popular shell with collectors and is sometimes taken for its commercial value. It has been suggested that the collection of so many shells was the cause of the outbreak of the Crown-of-thorns on the Great Barrier Reef because their removal freed the starfish from predator pressure

Above: A strange effect produced by marine weathering of rocks in Takapoto island in the Tuamotu Archipelago

Left: Two photographs of Long Key Underwater Reserve, set up in the waters off Florida by the United States Government to protect the coral formations

carnivores in any particular web than there are herbivores, so that the structure is like a pyramid. In all communities, on land or sea, the plants fulfil the role of primary producers; they alone can convert carbon dioxide and water into carbohydrates (in the form of sugars) and oxygen. They do this by the biochemical process of photosynthesis which uses their green pigment, chlorophyll, in the presence of sunlight.

Coral reefs are very complicated communities with a great variety of species taking part in the food webs. Corals themselves are not primary producers, and in some cases depend on the animal food that they can catch. They do, however, often contain microscopic plants in their tissues which are primary producers, and which in some cases may supply organic material to the polyp which houses them. Corals in their turn are preyed on by other animals which may depend wholly or partially on them as a source of food. A variety of organisms prey on corals, including both invertebrates like starfish, sea-urchins and sea-slugs and vertebrates in the form of fishes.

Without doubt the best known coral predator is the Crown-of-thorns starfish, *Acanthaster planci*, which has become notorious in recent years due to its apparent population explosion in parts of the Indo-west Pacific Region (a region including the Indian Ocean and the western Pacific). The genus *Acanthaster* is not represented in the Atlantic, so Atlantic corals in the Caribbean are not threatened by it. *Acanthaster planci* is a large starfish, and fully grown specimens may measure up to 40 cm (16 inches) across, or more. Unlike many starfish it has more than five arms or rays, and there are usually any number between 13 and 22. The whole of the upper surface is covered by quite large pointed spines which may reach up to 25 cm (10 inches) or more in length. The spines consist of a sharp, pointed skeletal rod, itself made of calcium carbonate, which is overlain by a thin layer of skin. The skin secretes mucus and venomous chemicals which are thus usually introduced into any wound caused by the spine. Very painful injuries have been reported by divers and bathers who have accidentally mishandled a specimen of *Acanthaster*.

Before 1963 this starfish was considered a rarity. It was first described in 1743, although it had been mentioned as early as 1705. Linnaeus named it in 1758 and subsequently a specimen, possibly collected during Captain Cook's voyage in H.M.S. *Endeavour* about 1770, was illustrated in a book published by Ellis and Solander in

Top: A small sand cay with characteristic vegetation of palms

Above: Coastline of Tahiti with its fringing reef

1788. Because this species of starfish was apparently so rare very little of its way of life, feeding and reproductive behaviour was observed or understood. During the Great Barrier Reef Expedition of 1928-1929 only one specimen was recorded despite a thorough investigation of several localities off the coast of Queensland.

In the early 1960s reports from the Great Barrier Reef region of Australia indicated an apparently unusual phenomenon. Large numbers of the Crown-of-thorns starfish were being encountered on some of the reefs in the north of Queensland and at other places in the Indo-west Pacific. In many cases the densities of the animals were remarkably high; some reports gave a figure of up to nine individuals per square metre (10·7 square feet) of reef. Similar reports continued to come in from various places in the Indo-west Pacific and new aggregations of individuals are still being brought to light at the time of writing.

The significant feature of *Acanthaster* is that it is a very efficient coral predator. Generally it inhabits quite shallow water and does not usually occur below the depths at which coral will grow. Its geographical distribution is limited by that of the coral which forms its diet. The Crown-of-thorns is normally regarded as a nocturnal animal which hides away in daylight in crevices between coral colonies. At night time it emerges to feed, and it does so by everting part of its stomach through the mouth which lies on its underside. Running along the underside of each ray of the starfish are the locomotory tube feet. These hydraulic organs can be extended by fluid pressure from within the animal's body and they are equipped with suckers at their tips to provide adhesion, and with muscles for making stepping movements. By means of these tube feet the

Above: A typical sand cay, surrounded by a wide expanse of reef flat. This picture is of Erskin Island, one of the Capricorn Islands, in the Great Barrier Reef Province

starfish can climb into various coral colonies and attack the polyps. When the stomach is everted it is closely applied to the live polyps and often wrapped round whole branches of colonies. The tube feet are often used to assist in placing the stomach folds in the correct position. When this has been accomplished digestion begins *in situ*, and so long as the starfish is left undisturbed the process lasts for several hours, up to six or even more. Then the stomach is withdrawn by special muscles, and the animal retires to a convenient hiding-place leaving behind it a whitened patch or feeding scar where the live polyps were. Such scars or patches are quite easy to see under water. A mature adult may consume up to 100 square centimetres (15 square inches) of coral polyps at a single feed, and it normally feeds every night.

A coral reef could support a small population of these starfish without a serious problem being encountered. When large numbers of this voracious carnivore are present, however, their food requirements cannot be satisfied by the natural growth and replacement of corals. In some cases very large areas of coral polyps may be killed and not replaced, so that, in turn, all the animals that depended on the coral for food or shelter will disappear. Understandably, many of the responsible authorities have been alarmed at the prospect of coral reefs in their areas being 'devastated' or 'denuded' by the activities of great numbers of *Acanthaster*, so there have been many calls for a programme of coral conservation, and for investigations into the biology of the Crown-of-thorns.

The principal object of the various investigations has been to find out what caused the population explosion in the first place. A number of people have suggested that pollution of the sea might be the cause of the apparent population outbursts, and significantly many of them have been reported as occurring near centres of human population.

Before the effect of pollution and other possible causes can be properly considered, however, there are other factors that should be taken into account. Firstly, most of the outbreaks of *Acanthaster planci* were discovered as a direct result of the comparatively new facility of scuba diving. Previous outbreaks may well have occurred, but they remain unrecorded because no one was in a position to notice them. Secondly, although most of the population outbursts were recorded near densely peopled areas, it must be remembered that it is precisely these regions that are most accessible to divers, and that vast areas of relatively inaccessible reefs exist where outbreaks might well have occurred which have never been monitored. The majority of scientists interested in the problem focused their attentions on the situation in Australia, but there the issue became confused because of the inevitable conflict between those wishing to exploit natural resources on the one hand and those advocating conservation on the other. The Great Barrier Reef Region of Australia is vast, and contains a wealth of natural mineral resources. Not surprisingly, the interests of industry were directed towards these, while conservationsists were quick to argue that the influence of man, particularly his industrialization, might affect the balance of the ecosystem. As a direct consequence, various research programmes were initiated with differing points in view. These included increasing our knowledge of the general biology of *Acanthaster planci*; looking for ways in which pollution or other factors could affect its

population; looking at the effect of the starfish on corals and how they might recover; and investigating possible methods for the control of *Acanthaster*.

The investigations on the general biology of the Crown-of-thorns starfish have provided a great deal of information on its way of life, especially on subjects like food selection, feeding behaviour and reproduction. Recent research indicates that *Acanthaster* is distributed generally over the Indo-west Pacific region wherever the appropriate shallow-water habitats exist and the prevailing temperatures permit the growth of reef corals. In a normal situation under these conditions, it appears that *Acanthaster* is found in quite low population densities: between five and twenty individuals per kilometre (0·6 miles) of reef face. Under such circumstances individual starfish are generally surrounded by a great variety of coral as potential food, and they are able to select particular species. That they do indeed choose like this is evident from investigations on the feeding scars which are left on the coral. There is some evidence, too, that once accustomed to a particular diet, various species of starfish will continue to feed on the same type of food although other alternative sources may be available. In *Acanthaster*, exactly what determines this preference for a particular species of coral is still not understood, but it seems likely that the overall shape and form of the colony, whether or not it is positioned in exposed or sheltered places, whether or not it is surrounded by rocks or sand, and how powerful its stinging cells are could all be important. Corals that have been noted as forming the diet of *Acanthaster* include branching forms like some species of *Acropora* and more rounded types like *Favia* and

Below right: A view of Yap Island, in the Caroline Islands. The outer edge of the fringing reef is shown by the breaking waves

Bottom right: A group of islands in the archipelago off San Blas in Panama, surrounded by a reef

Below: Coral islands in the Indian Ocean with fringing reefs. The varying depth and substratum of the lagoon can be seen from the different colours

Goniastrea. A shortage of preferred types, of course, leads inevitably to *Acanthaster* extending its diet to include less favoured species.

Where slightly higher population densities have been recorded, such as from 30 to 200 individuals per kilometre (0·6 miles) of reef face, a slightly different distribution occurs. In such situations the individuals may form small groups numbering between three and thirteen. These groups may have come together as a result of an attraction to particular colonies of their preferred coralline food, especially if the general availability of such corals has decreased in the area due to a prolonged period of predation by other Crown-of-thorns starfish. In addition, other factors such as reef topography, exposure and quality of substratum may channel individual starfish towards each other, since *Acanthaster* avoids sandy bottoms when rocks are present, and shows a preference for places which provide some shelter from wave action yet are not too calm and hot like some lagoons. In these small groups of starfish the operation of feeding seems to be synchronized and, generally speaking, where the corals are large enough the members of each group begin to feed each night at the point where they left off on the previous night. When the coral colony has been finally exhausted the group may break up to go in search of other suitable corals. Experiments carried out in the Red Sea have shown that not only can a starfish distinguish between different species of coral by scenting them in a constructed maze, but that an *Acanthaster* will find other feeding individuals more attractive than non-feeding ones, and this is evidently another reason for the formation of these small groups.

Under certain circumstances, as yet not fully

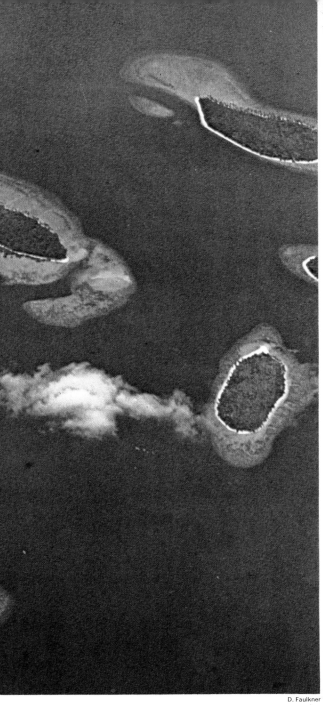

Left: Aerial view of a group of small coral islands in Indonesia. The photograph clearly shows the extent of the reef around each one and the distribution of the deeper (dark blue) water

Below: The shore of an island in the Visayan Sea in the Philippines. It consists largely of coral debris

such circumstances is not clear, but it could be that they disperse, moving into deeper water in an attempt to reach other reefs.

For an animal like *Acanthaster*, which at spawning time merely sheds its eggs and sperms into the surrounding sea-water, life in aggregations would seem to be advantageous since the chances of successful fertilization of the eggs must be increased. If the population density is very low such chances must be slight. A variety of other marine invertebrates are known to swarm at spawning time, but it still remains to be demonstrated that this is the reason for the 'plagues' of *Acanthaster*. It seems more likely that the mutual attraction of feeding individuals, coupled with the influence of the reef topography, would bring about the start of an aggregation—beginning with the small groups that have been described.

Another interesting feature of the life of *Acanthaster planci* derives from the fact that in common with a great number of other invertebrates dwelling on coral reefs, including the corals themselves, it has a life-cycle which involves a drifting, planktonic larva. This is moved about by currents, and feeds constantly on minute drifting plants, until the time comes for settlement and metamorphosis. Just before metamorphosis the larva changes its behaviour and sinks to the bottom to explore suitable surfaces. At this moment one of its chief predators would appear to be the coral polyps themselves, and work in laboratories has shown that settling larvae of *Acanthaster* may be quickly overcome by the polyps of certain species of *Acropora*, including those species which become food for the adult starfish. Consequently it is suggested that the

understood, *Acanthaster* occurs at much higher densities than those described so far, and concentrations of up to 500 or more individuals within 100 m (328 feet) of reef face have been recorded. The number of examples of such extraordinary concentrations which have been accurately described by scientists, however, is probably smaller than is sometimes supposed. Under such circumstances the behaviour of individuals has been seen to differ considerably from that described so far. Probably due to the inevitable shortage of food which must accompany a population increase, the starfish have been seen feeding, in the open, by daylight. In a number of the cases that have been studied, these so-called plagues have been seen to dwindle once the greater part of the live coral has been exhausted, leaving behind a mass of coral skeletons. Where all the starfish might have gone under

Left: The coastline of Felidu Atoll in the Maldive Islands

Below left: A stretch of beach on Zanzibar; the island is surrounded by coral formations which are particularly large along the western coast

feeding activities of adult starfish may prepare suitable areas for larval metamorphosis in the form of dead coral skeletons. It has been claimed that the dead skeletons of *Acropora* are therefore attractive to the larvae. This somewhat unusual situation, where the prey of the adult starfish would appear itself to be a major predator of the starfish larvae at their moment of settlement, probably means that an already growing population of starfish is able to recruit new members in greater numbers than usual from the plankton. The closer the adults are together, then the better are the chances of successful fertilization. Interestingly, such a theory is supported by the observation of large numbers of *Acanthaster* on coral reefs which have been blasted and dredged.

These theories are not, however, the only explanations put forward to account for the appearance of *Acanthaster* in large numbers. Other explanations which have been suggested differ in that they assume the activities of man to have been in some way the initial cause. Whilst this may be so, it must be remembered that swarms of land animals, such as the periodic activities of the Norwegian lemming are not due to man's influence, and the same could be true of *Acanthaster*. One explanation for the Crown-of-thorns problem based on man's activities suggested that the activities of shell collectors were to blame in the Great Barrier Reef Province. The shell of the Great triton or Trumpet triton, *Charonia tritonis*, has long been popular with collectors and great numbers of this large, predatory whelk have been taken; the shells of fully grown adults may reach 40 cm (16 inches) or more in length. *Charonia* feeds on a variety of invertebrates including *Acanthaster planci*, and in the last few decades so many thousands have been removed that, according to this theory, their removal freed the Crown-of-thorns from the pressure of predators, so that its numbers were allowed to increase. Further research has shown, however, that this is unlikely to be a sufficient explanation, since *Charonia* accepts a variety of food and does not prey on *Acanthaster* alone. Furthermore, in the Red Sea, where the whelk appears to be scarce, there is generally a low population of *Acanthaster*. No other known predator of the adult starfish is thought to have been interfered with by man's activities. Whether or not this is also true for predators of the Crown-of-thorns larva cannot be decided as so little is known of its planktonic life. But it is quite conceivable that the effects of chemical effluent, oil spillages and so on are to blame for the elimination of some controlling factor in the larval phase.

A great deal has been said here about the Crown-of-thorns starfish because of the apparent severity of the problem, but it is, of course, only one of many predators of corals. The Crown-of-thorns issue does serve to illustrate very well many aspects of the ecology of coral reefs and the startling way in which they can deteriorate if not protected. Among the other invertebrates which attack coral polyps is the sea-urchin, *Diadema setosum*, known for its long slender poisonous spines. It rasps a variety of organisms, including corals, with its teeth and ingests the soft tissues. A variety of fish feed on corals; some appear to chew off portions of coral colonies with their powerful teeth, and in so doing they take a great deal of calcium carbonate into their bodies which passes out with the faeces. Others, like the large Donkey fish in the Indian Ocean, have enormous bony humps on their heads. By swimming fairly quickly into a stand of branching corals they can break off pieces which they can then feed on.

Apart from the organisms which actually attack the coral polyps a great number attack the limestone of the reef itself. They either bore into it like some sponges, bivalves and sea-urchins or they scrape it away as they browse on organisms that grow on it.

Coral reefs exist as dynamic communities which embrace many specialized organisms each occupying a particular place or niche in the community, and each helping to establish or break down that community. While the balance between the building and destroying components is maintained, reefs will continue to flourish and provide magnificent underwater spectacles. When one element gets out of control, as has been shown by the Crown-of-thorns starfish, the community is quickly destroyed. Although it seems likely that the so-called 'plagues' of starfish are naturally occurring phenomena, they should serve as a timely reminder of what could happen if man's activities in the region of corals and coral reefs are not monitored and controlled.

The coral environment

For many generations corals were regarded as plants. It was not until 1726 that the French naturalist Peyssonnel showed that they were animals. He realized that inside each minute limestone cup or theca there was an animal which could respond to stimulation by making a movement. Normally plants can respond to stimuli only by growth changes.

The word coral was probably first used to describe the semi-precious red coral known since classical times from the waters of the Mediterranean and the Atlantic. Today the term usually refers to those members of the phylum, or division of the animal kingdom, known as Cnidaria which have skeletons made from calcium carbonate, but it is especially applied to the true or stony corals called madrepores which build reefs. All members of the phylum Cnidaria are characterized by the possession of stinging cells or cnida. These cells are used for the immobilization of prey and for defence of the animals which bear them. Many cnidarians, such as the jellyfishes and the sea-anemones, do not secrete a chalky skeleton. The cnidarians are sub-divided into three classes: the Hydrozoa (which includes the freshwater polyp *Hydra*, the sea-fir *Obelia* and certain other organisms which have been called corals); the Scyphozoa (jellyfishes) and the Anthozoa (which includes the sea-anemones, soft corals and the true or stony corals).

A number of hydrozoans have some sort of skeleton and in many cases it is composed of calcium carbonate. One example is the so-called fire-coral, *Millepora*. Millepores do in fact contribute to the formation of reefs because they secrete quite substantial plate-like skeletons, but they are not able to form reefs by themselves.

Left: The delicate ramifications of a live sea-fan. The tiny polyps can be seen as they feed. These animals grow in colonies which generally branch in one plane

They are known as fire-corals because their defensive polyps are equipped with powerful stinging cells which can cause considerable pain when they are touched.

There are no coral-like polyps in the class Scyphozoa. In the class Anthozoa, however, there are various forms of polyps which secrete chalky or horny skeletons as well as others like the sea-anemones which do not. Anthozoan polyps can be distinguished from Hydrozoan ones because their anatomy is more complicated. One important distinguishing feature is the distribution of flaps of tissue called mesenteries or septae which are arranged vertically around the interior of the stomach, so dividing it into a number of pouches. In most advanced Anthozoans these flaps are not evenly distributed in all planes round the gastric cavity, but tend to be arranged in opposite or complementary halves, so that when a cross-section of the polyp is examined it is seen to display an internal symmetry that is not radial but is reminiscent of the bilateral symmetry of the higher animals. Further differences separate the two classes. Almost all hydrozoans have a free-swimming jellyfish stage in their life-cycle, but this stage is missing from the life-cycles of the stony corals and other anthozoans.

The members of the class Anthozoa are themselves divided into two major groups. One of them, the group Octocorallia, includes all those polyps with eight branched tentacles. The presence of branched tentacles immediately distinguishes the Octocorallia from almost all the other anthozoans. The Octocorallia is itself divided into a number of orders of which two should be mentioned in more detail as their members are sometimes called corals. The first is the order Alcyonacea, popularly called soft corals. The common name has arisen because, with a few important exceptions, these cnidarians do not secrete hard, rigid external skeletons. Rather they form within their tissues a great number of minute chalky spicules which are loosely scattered throughout the colony, giving it a flexibile texture. There are alcyonarians which are not soft, however, one being the organ-pipe coral, *Tubipora musica,* where the polyps secrete hard tubular red skeletons. Moderate-sized colonies of this species occur on many Indo-west Pacific reefs.

The second order of the Octocorallia which should be mentioned is the Gorgonacea. These organisms generally form erect, branching colonies which often appear as fan-like or plant-like growths. The polyps secrete a central skeleton, common to all of them, which runs along the middle of the colony's branches. This is composed of calcium carbonate and a protein called gorgonin. To this group belong the sea-fans and sea-whips, many of which branch in one plane only, as well as organisms like *Corallium rubrum,* the well known semi-precious Red Coral from the Mediterranean. Although it is so often called a coral, it must be emphasized that this popular Red Coral is not a close relative of the true or stony corals of the tropical reefs. The Mediterranean fauna does include several species of true corals, but the physical conditions are not suitable for reef formation.

The second major group of the Anthozoa is the Hexacorallia. This includes all the anthozoans with more or less than eight tentacles per polyp, and in almost every case they have unbranched tentacles. In many cases the polyps are six sided, but it may not always be easy to see this. The Hexacorallia includes a number of orders. Some of these like the Actiniaria (sea-anemones) are familiar to many people, but others like the Antipatharia (so-called black corals) are rather obscure even though they are found in most oceans of the world. Black corals grow as bushy colonies which secrete a blackish skeleton of horny material. They are not important in reef-building and must be carefully distinguished from the few true or stony corals in the order Madreporaria which also have black skeletons. In the antipatharians, the skeleton forms a central supporting strand which is cloaked in soft tissue with the polyps on the outside, and this structure is entirely different from that of the cup-like skeletons of the madreporarians. The group Hexacorallia also includes orders of anemone-like polyps which, though they are of great

F. Quilici

Above: Alcyonarians belonging to the genus Sarcophyton on the Great Barrier Reef. In the foreground is a typically folded colony with open polyps; in the background a similar colony with retracted polyps

Above left: A group of non reef-building (ahermatypic) coral polyps. The extended tentacles and slit shaped mouths can be clearly seen

Left: A diver surfaces with two branches of soft coral

interest to zoologists, are not important in the study of coral reefs.

The anatomy of the true or stony coral polyps has been described briefly but for a proper appreciation of the polyp's interaction with its environment a closer examination is essential. Each polyp consists essentially of a sac-like body with a mouth at one end. The body wall is made of two layers of cells known technically as ectoderm (on the outside) and endoderm (on the inside). The top of the body is disc-like, and at the edge of the disc are one or more rings of tentacles. These organs are hollow and the space inside them communicates with the inside of the sac-like body. The tentacles are responsible for trapping the polyp's food, and they are assisted by many stinging cells or cnida which are usually grouped in batteries on their outer surface. The mouth itself is frequently slit-like, or at least not perfectly round. It is surrounded by muscles to open or close it.

The mouth communicates with the interior of the sac-like body by way of a tubular sleeve which hangs down some way inside. This is the pharynx, and through it any food must pass on its way to the stomach. Since there is no anus, the mouth and pharynx are also responsible for removing food waste. The stomach itself is formed from the cavity inside the sac. It is not a simple cavity, however, but is divided up into a number of pouches or sectors by the flap-like septa which run vertically up and down the polyp linking the disc which surrounds the mouth with the bottom of the polyp. The function of the septa has been the subject of some speculation. Some authorities consider that they increase the surface area of the stomach and thus the area which is available for absorption of digested food. A similar role has been proposed for the thread-like gastric filaments which branch off from the bases of the septa. Other authorities believe that the septa have been developed to accommodate muscles which run from the base of the polyp to the disc and which are embedded in them. These muscles are responsible for withdrawing the tentacles and mouth disc when the polyp contracts. When this contraction takes place some of the fluid within the stomach has to be expelled via the pharynx and mouth.

The lower regions of the outside of the body are particularly concerned with the relationship of the polyp to the skeleton of calcium carbonate in which it is supported. In a dead piece of coral, the bottom of this cup, known technically as the theca, can be seen to be patterned by a number of delicate plates, arranged vertically on their thin edges, and radiating from a centre point. These

Above: Carvings made of precious coral, Corallium rubrum

Left: A delicate branching sea-fan (gorgonian) showing the typical two-dimensional growth pattern with polyps open for feeding

Above right: Section of a coral polyp. For the sake of simplicity the soft septa and gastric filaments inside the stomach have been omitted. Adjacent polyps are connected by canals linking their gastric systems

Right: The sea-fan, Eunicella verrucosa, from European waters

Far right: A close-up of the small polyps of another sea-fan. The eight branched tentacles which surround the mouth can be seen on each polyp

are known as sclerosepta, and have been secreted from the base of the polyp which is not flat and disc-like as are the bases of sea-anemones, but is wrapped closely over the sclerosepta.

The depth of the theca varies considerably according to the species in question. In some madrepores the polyp can be completely withdrawn inside it, but in others this is not possible and much of the polyp will be exposed to view regardless of whether or not its tentacles are expanded.

The two cell layers of the body wall, ectoderm and endoderm, are separated by a thin layer of jelly called the mesogloea. It is the presence of this jelly in far greater quantities which gives the jellyfishes, relatives of the corals, their characteristic gelatinous texture. The ectoderm contains a variety of cell types including muscle cells, nerve cells, secretory cells and cnida or stinging cells.

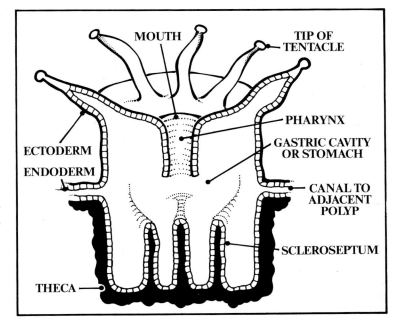

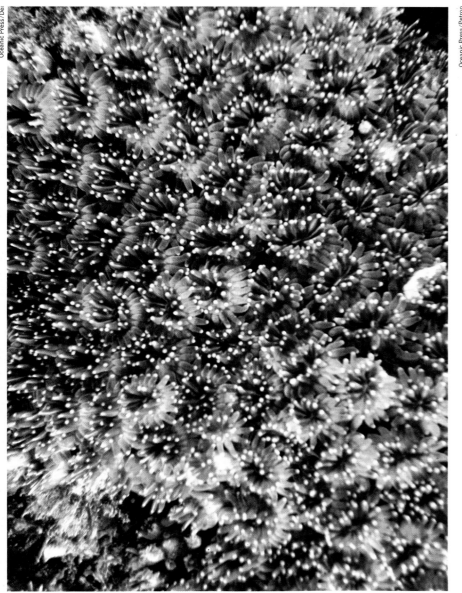

*Above: A colony of reef-building coral (*Galaxea *species)*

Above right: Branch of Acropora *species which often form bushy colonies*

Left: Acropora *on the sea-bed in the Palau Islands. This coral is often inhabited by brightly coloured damsel fish. The fish shown here are a species of* Chromis

The cnida are of great significance in the life of a coral polyp, and normally several varieties occur in one animal. When an appropriate stimulus is applied either by touch or in the form of certain chemicals, they discharge a minute hollow thread. Prior to discharge this thread is stored coiled within the cnidum cell, and at discharge the thread turns inside out and protrudes from the cell into the water. Some of these threads are barbed and can penetrate the bodies of other small organisms, some are open ended and can inject venom into the organism that they have penetrated, and others are twisted so that they form a coil which ensnares small structures or minute swimming animals. The combination of various types of cnida on the tentacles and outer surfaces of the body give the polyps the capacity to trap prey so that it can be manipulated by the tentacles and passed to the mouth.

The endoderm of the reef-building stony corals, which are technically known as hermatypic corals, may be distinguished from that on non reef-building, or ahermatypic, corals by the possession of special carrier cells which contain minute unicellular plants known as zooxanthellae. The function of the zooxanthellae has been the subject of great interest for some time. Recent work on coral physiology and biochemistry has shown that in some hermatypic species these zooxanthellae may provide the polyps with nutrients as a result of their photosynthesis. The degree to which this occurs throughout the true corals is not yet known, but it is now certain that some coral polyps are not merely dependent on the food they can catch with their tentacles as had formerly been supposed. It seems likely that the zooxanthellae also assist in the formation of calcium carbonate by absorbing carbon dioxide and bicarbonate.

Those stony corals which have long tentacles are able to catch small swimming organisms by means of their stinging cells, and the tentacles can then push the prey towards the mouth. Most polyps appear to be very sensitive to chemicals released from prey and their detection initiates the feeding responses. In some species irregular contractions of the disc surrounding the mouth serve to bring the opening towards the food, thus speeding its ingestion. During or before swallowing, the gastric filaments, which grow from the bottom of the septae, may be partially extruded through the mouth. Smaller polyps may depend more on trapping fragments of food, which fall on to the colony, in a web of mucus which is secreted by the ectoderm. This trapped food is then swept towards the mouth by means of currents created by the cilia, and this action reveals some interesting distinctions in polyp behaviour. In almost all coral polyps the ectodermal cilia of the disc and tentacles sweep material away from the mouth and beyond the periphery of the polyp; presumably, this assists the prevention of clogging by silt in the exposed parts of the polyps. In those polyps which depend on mucus and ciliary beating for feeding, the ciliary beat reverses its direction when food is present, but, once the food has been ingested and the stimulus disappears, the cilia revert to their normal cleansing role. There are also cilia in the tubular pharynx of the polyp which are responsible for wafting small particles of food into the stomach.

Most reef-building madrepores retract their tentacles in the day and extend them at night when the animals of the plankton rise towards the surface of the sea and thus come within their reach. Recent work has shown that a coral reef is

T. Poggio

Below: Coral colonies on the reef crest showing at low tide

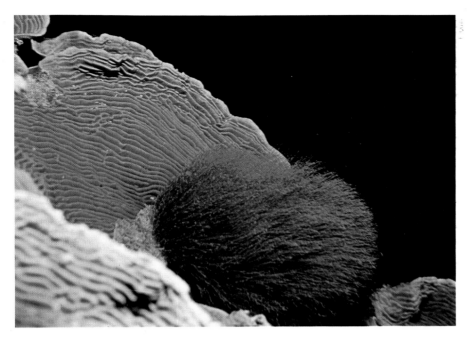

capable of removing about 91 per cent of the phytoplankton (plants) and about 60 per cent of the zooplankton (animals) from the sea-water passing over it. The corals themselves, however, account for only part of this consumption since there is a rich variety of associated organisms which are also straining the sea-water for plant and animal food.

Clearly the shape and form of a coral reef depends to a great extent on the nature of the coral colonies which have formed it and the manner in which they have been influenced by the environment surrounding the reef including the shape and slope of the sea-bed on which it stands. Evidently, the most fundamental factors in the form of a colony are the number and type of coral species available to colonize and shape it. In the tropical Atlantic, the diversity of coral species is not very great, and the number which

Above: Plate-like colonies of encrusting corals and clumps of green algae

Right: The famous precious coral (Corallium rubrum) and solitary polyps of true corals in the Mediterranean. The reddish-brown and white encrustations are sponges

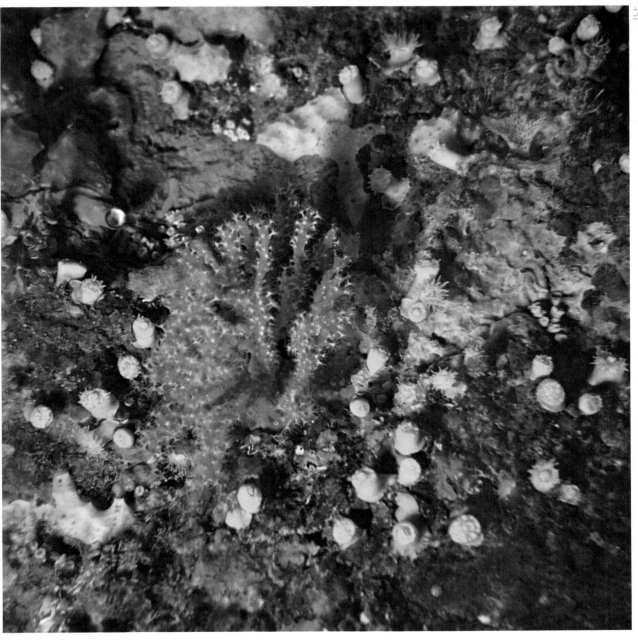

Left: A blue surgeon fish swimming in front of the skeleton of a dead sea-fan

*Above and right: Two photographs showing different growth patterns of fire-corals (*Millepora*), one branching and one plate-like*

*Below: Growths of Stagshorn (*Acropora *species) and surgeon fish off San Blas Islands, Panama*

Above: The true coral Goniopora *with its polyps fully extended*

significantly contribute to the formation of reefs may be rather less than 50. In the Indo-west Pacific region, however, recent estimates of the number of reef-building species give a figure in excess of 500, and as many as 200 species may occur on one reef alone. One genus, *Acropora,* is represented in the Caribbean region by only three species, but, when it is put together with a closely related genus, *Montipora,* the two genera are represented by about 250 species in the Indo-west Pacific. The Atlantic corals are effectively isolated from their close relatives in the Indian and Pacific Oceans. They are cut off to East and West by the land masses of Africa and the Americas, and to the North and South by cold waters through which their planulae could not pass without perishing. This isolation has led to the development of reefs in the tropical Atlantic which do not show the community structures typical of the Indo-west Pacific The richest coral associations in the world occur at places like the north-eastern coasts of Australia where the

Right: The unusual appearance of a single polyp of Fungia actiniformis, *a non reef-building coral. The granular appearance of the surfaces of the tentacles may be due to batteries of cnida*

Right: A very distinctive 'coral' belonging to the Hydrozoa, many of which do not build calcareous skeletons. The skeletons of this group are not important in reef-building

Indian Ocean and Pacific Ocean communities meet. At other places like the Red Sea, the local coral fauna is flourishing, but not so varied.

Reefs usually flourish on the eastern seaboards of islands and continents, because the circulation of currents in the western regions of seas and oceans creates preferable conditions for reef development. Appropriate conditions seldom prevail along the western coasts of land masses.

Reefs themselves occur in a variety of forms. Darwin, in his book *Coral Reefs* first published in 1842, grouped all reefs into three principal classes: atolls, barrier reefs and fringing reefs.

Atolls are characteristically formed in mid-ocean, far removed from land masses. Darwin noted that they were generally surrounded by deep water on all sides, but that they enclosed a shallow lagoon where the physical conditions of the water showed a marked contrast with those present in the ocean outside the ring-shaped reefs. In some cases the tops of these reefs protruded from the sea and had sparse vegetation.

43

Barrier reefs, as described by Darwin, occurred in a variety of shapes and sizes. In many cases he noted that they encircled rocky islands, and that the distances separating the barriers from the islands or land masses that they protected ranged from a mile or so to hundreds of miles. The reefs generally enclosed a relatively shallow area of water analagous to the lagoon in the centre of an atoll. Alternatively, barrier reefs as Darwin saw them could occur as strip-like or chain-like reefs in mid-ocean. The significant characteristic was that they stood, like atolls, in deep water.

In contrast, fringing reefs occur in close association with the shore and their horizontal width is directly related to the angle of slope of the seabed on which they stand. Where this is sharp they tend to be narrow, and where it is gradual they tend to be wider. Generally speaking, however, they are seldom very wide and few islets can therefore form on them. Unlike barrier reefs and atolls they occur in relatively shallow water. Darwin's definitions proved popular for a long time, but quite recently it has become apparent that they are not necessarily adequate to deal with a complex situation such as exists in the Great Barrier Reef Province of Australia.

The Great Barrier Reef is situated on the continental shelf that runs along the eastern seaboard of Australia and continues along the coasts of New Guinea. It is an immense chain composed of a great variety of reef forms which stretches for over 1,900 km (1,200 miles) and encloses a total area in the region of 260,000 square kilometres (100,000 square miles). For the greater part, the water where the reefs stand does not get deeper than about 150 m (500 feet) and it is often shallower. For this reason it was recently suggested that the term 'barrier', as used by

Below: A brain coral. The outline of each irregular-shaped polyp is marked by the position of the red retracted tentacles. Taken as a whole the colony bears a remarkable resemblance to a mammalian brain

I.C.P.

Right: Transect across a hypothetical island's shore and reef to show the principal zones

Below: Close-up of a brain coral showing the green retracted tentacles, the grey skeletal margins of the polyps and their interlocking shapes

Below right: By contrast with brain corals, the retracted polyps of this coral show simple regular outlines

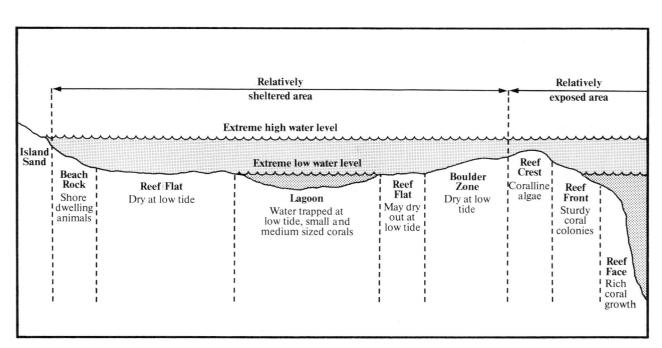

Darwin, was inapplicable to the reefs off the coast of Queensland Australia, and that in many ways these reefs did not resemble the deep-water barrier reefs of the central Indian and Pacific Oceans. The term Great Barrier Reef Province has therefore been used to describe the area formerly known as the Great Barrier Reef. It is divided, for scientific purposes, into three regions. The northern one includes all those reefs to the north of latitude 16°S. where they stand in water not much deeper than 40 m (130 feet). The central region lies to the south of this and reaches down to latitude 21°S. where depths range from 40 m to 60 m (130 to 200 feet). In the southern region the water reaches a maximum depth of around 150 m (500 feet). The outer edge of the continental shelf is marked by an almost continuous chain of reefs, which at their northernmost extremity lie close to the land, but further south, as the edge of the continental shelf stretches further and further away from the land, the outer 'barrier' as it is still generally termed becomes increasingly fragmented. Within the area enclosed by the outermost reefs and the shores of Queensland lies a great variety of other reefs, both small and large, low wooded islands, and continental islands. The composition of a typical coral community can be seen by examining a transect at low tide from the shore to the reef edge on one of these low wooded islands. A transect is a record of the organisms and communities along a straight line, and gives a good indication of the zonation in any locality.

The island sand runs down to roughly the level of extreme high tide. On some sand cays, where vegetation has developed and stabilized the sand, one usually encounters so-called 'beach rock'. Beach rock is formed from coral sand, shells and

Right: The soft coral Sarcophyton *showing the numerous extended polyps each with eight tentacles*

Below right: Close-up of branches of Acropora *species showing the individual skeletal-cups with the polyps withdrawn. Those on the tips of the branches are larger and coloured yellow, whereas those on the sides of the branches grow with a typical scoop shape*

Below: Different species of coral and a sea-fan, on the sea-bed off the Palau Islands

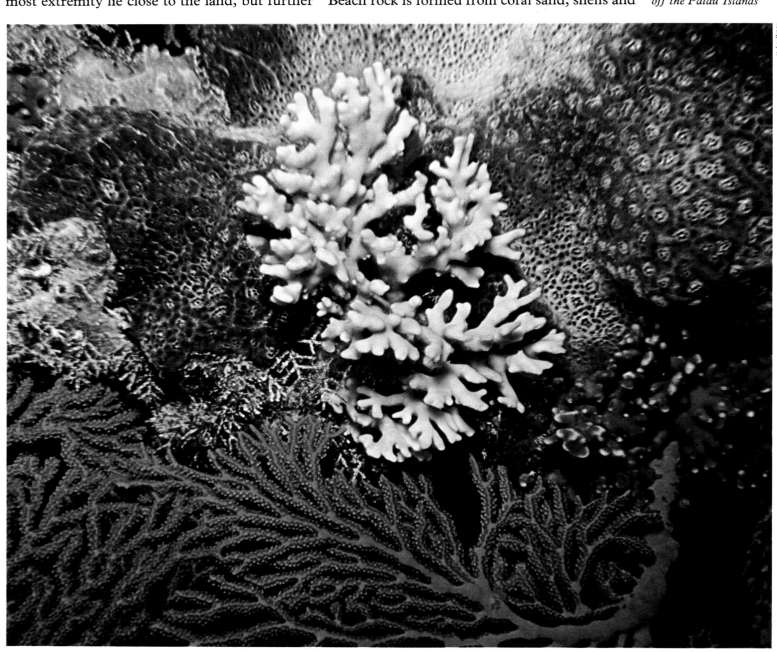

48

Far left: Three fan-worms. They live within colonies of true coral and the double fans on their heads are used to collect food and oxygen

Left: A group of sea-squirts growing on a reef. The large apertures are the openings through which water is inhaled bringing food and oxygen. Water and waste then leave via the smaller openings

Below left: The reef flat at low tide

Right: A close-up view of Fungia *which is a solitary coral. It can reach a diameter of 25 cm (1 foot). The tentacles are withdrawn and the mouth lies in the central depression*

Below: Group of colonial sea-anemones on dead coral

other objects originating from the reef; it is the result of chemical action dependent on the influence of the tides, the sun, rain and pressure. Obviously, the more recent the formation of a coral cay, the less beach rock there is likely to be. Beach rock usually occurs as a band or strip running parallel to the shore line, and its upper limit is towards the high tide level. In many cases it occurs only on the more exposed sides of the islands, but sometimes it runs all round their circumference, simply being best developed on the most exposed side.

The beach rock zone can be divided into three sections: in the uppermost the rock is very soft, but farther towards the sea a much harder region is encountered with sharp, jagged formations, while the rock in the third zone is smooth and slippery, often being covered with algal growths.

The bottom zone of beach rock usually adjoins the so-called 'reef flat'. This is an almost level area which slopes, if at all, very gently towards the edge of the reef. The form of the reef flat can vary from reef to reef. In some cases it consists of a homogeneous area of coral rubble covered with encrusting algae. This rubble is made from pieces of broken coral which have been washed over the top of the reef on to the flat, and which have then provided the substratum for algal

Left: A large sea-anemone which is almost completely closed. The club-tipped tentacles of another anemone can be seen beyond. The black and white Domino damsel fish is often associated with these anemones

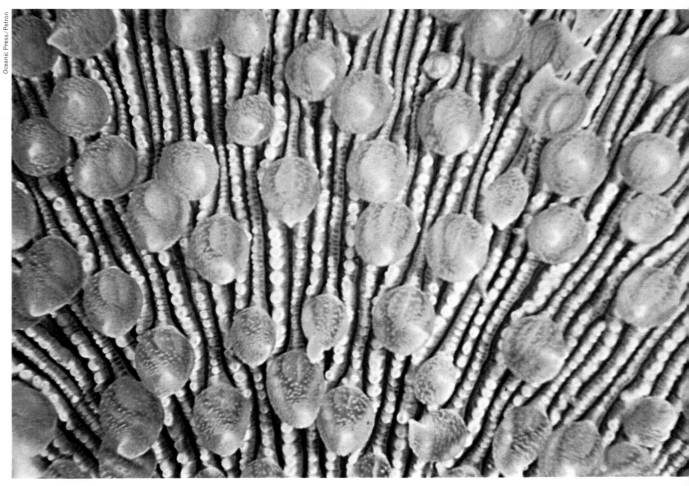

Right: Enlarged detail of a Fungia *polyp in which one can clearly see the wafer-like scelerosepta and partially extended tentacles*

Below: Many small fish find food and refuge among branches of Acropora

development. Red, brown and green varieties of algae may be found inhabiting this region, and this flora provides a rich feeding ground, both for the invertebrate animals which dwell among the rubble and for fish which come in with the tide. Not all reef flats consist merely of homogeneous coral rubble, however. There are instances where the flat dips down and rises again towards the reef edge, so that at low tide there are standing pools of sea-water or lagoons. These areas may be carpeted with a sandy bottom, and are frequently populated by sea-grasses like *Thalassia*, which are relatives of the land grasses, and their strong roots and stems may become the habitat for a variety of invertebrates and algae. The presence of this grass also helps to bind the sand and to protect it against the effects of wind and tide.

When the tide is out, the casual observer walking across the algal rubble during the daytime would see little evidence of the animals which populate it, since most of them are burrowers and crevice dwellers. A list of species would comprise a host of shrimp-like animals, worms, brittlestars and molluscs.

Wherever water is retained on the reef flat coral colonies can occur. One species characteristic of the reef flat is *Porites lutea*. This develops as rounded colonies which may be a metre (3 feet) or

Above: A stunning juxtaposition of colour and form. Black feather-stars perching on a sea-fan while they filter the sea-water for food

more in circumference, but owing to the shallow water at low tide they rarely grow to any great height, and the tops of the colonies are killed by exposure to air. Thus the colony can only grow outwards and increase its diameter, and in so doing it is in a sense reflecting atoll formation; the flat top holding water in a saucer-like depression which parallels the shallow lagoon within the atoll. Hence these colonies are sometimes described as micro-atolls, and they have recently become objects of interest to reef researchers.

Beyond these pools of trapped water, when they are present, the reef flat surface may rise very slightly towards the crest of the reef itself. So long as there is sufficient water to cover them, this area is characterized by flourishing growths of corals which are not able to withstand the more exposed positions on the seaward side of the reef. Between these richly growing coral colonies live a great variety of small fishes and invertebrates. Normally, lying between this community and the reef crest, the highest point of the reef proper, is a relatively barren area which dries out at low tide and is characterized by a number of boulders and dead shells. This is known as the boulder zone. It is not generally populated by many animals, and its extent depends largely on the conditions prevailing on and around the reef.

Beyond the boulder zone, if there is one, is the reef crest itself. This is the most conspicuous feature of many reefs. It is the area of maximum exposure, where the waves have forced the surface smooth and free of debris. The principal inhabitants of this zone are the coralline algae or nullipores as they are sometimes known. They encrust the rocks and provide a very hard, knobbly surface, which is generally pink or purple due to the pigments these remarkable

Above: During the breeding season, gulls like this can provide islanders with a steady supply of eggs

plants contain. These plants extract calcium carbonate from the surrounding sea-water—as do the corals—and use it to form tough branching growths which in some cases even resemble coral. The coralline algae are not restricted to the reef crest, but form an important section of the entire reef community; they both bind the corals together and contribute to the substance of the reef. They are principally represented on the reef crest by species of three genera *Lithothamnion, Lithophyllum* and *Porolithon*. Few animals can tolerate the severe conditions which normally prevail on the reef crest.

Beyond the reef crest the reef begins to slope gradually down towards the sea-bed. The extent of this gradual slope varies from reef to reef, and at different points along any one reef. Generally the area is characterized by flat platforms of coral rock and sturdy, short branching coral colonies like Stagshorn and some species of *Pocillopora* and *Montipora*. Coralline algae occur in this region too, and occasionally more massive coral colonies. At high tide this area is populated by a variety of brightly coloured fishes which move up from the lower regions. The invertebrates which occur here are specialized for life in very turbulent conditions. They must be able to feed whilst at the same time not running the risk of becoming dislodged by the waves.

At the outer edge of this seaward slope the angle of dip may dramatically change, and the reef's profile abruptly drop to the sea-bed which may be some way below. This is the reef face, the area of strongest coral growth. Those species of coral which can withstand the exposure are able to form rich growths in the clear, highly oxygenated water which brings them planktonic food from the open sea.

Diving in coral seas

For a diver the first impression of life in a coral sea can be an almost overwhelming experience, particularly if he is fortunate enough to choose a spot rich in reef life. The Red Sea, for instance, situated between the vast deserts of north-eastern Africa and Arabia, provides an area of outstanding natural beauty. The great beauty of life under the water, rich in colour and activity, contrasts markedly with the barrenness of the desert which reaches down to the water's edge.

The Red Sea can be said to be peculiar for several reasons. Firstly in geological terms it is a comparatively recent sea having probably been formed only about ten million years ago at the end of the Pliocene epoch. Secondly, it was formed from a drowned rift valley so that it shelves quite steeply down to a maximum recorded depth of 2,800 m (9,200 feet) and has no continental shelf. The sea-bed at this depth is strange and poorly understood, but appears to be rich in mineral wealth. Because the Red Sea lacks a continental shelf on which reefs can form as they do, for example, off the coast of Queensland, Australia, there is only, besides the typical fringing reefs along its shoreline, an irregular pattern of off-shore reefs. These reefs grow on raised submarine banks wherever they happen to be close enough to the surface for coral to develop. In the central and southern regions of the Red Sea some of these banks have developed in such a way as to form islets like those which make up the Suakin and Dahlak Archipelagos. Thirdly, the physical conditions in this part of the world make the physical characteristics of the sea different from most others. The average air temperature ranges from 26°C (79°F) in winter to 32°C (90°F) in summer, and that of the surface of the sea from 25°C (77°F) to 30°C (86°F) at the same times of the year. The air temperature in summer can reach the incredible level of 45°C (113°F) on occasions. This extreme heat causes a great evaporation from the sea's surface. The water lost in this way is replaced not by permanent rivers which flow into the sea, but to a great extent by a flow of sea-water which enters at its southern end via the Gulf of Aden, bringing with it more salt. Consequently the salinity is high, especially near the surface. The lack of any regular fresh water run off from the land also means that the underwater visibility is generally very good and that the sea is not cloudy but very clear and free from suspended matter. There are small tides at the northern and southern ends of the Red Sea, but in the central region the sea is virtually tideless. All these factors interact to produce conditions peculiar to this sea.

The open waters of the Red Sea are not themselves particularly productive, unlike the reefs and lagoons which swarm with life. This is due to the depth and the effect of the physical conditions surrounding the Red Sea on the sea-water itself. The lack of permanent rivers entering the sea means that the supply of minerals is almost non-existent, and the high temperatures prevailing all the year round ensure that the surface waters remain hot and thus greatly isolated from the cooler waters below—a phenomenon known as the thermocline. The thermocline is the interface between water masses of different temperatures and takes the form of a

Left: A variety of echinoderms may be encountered on a coral reef. This is a cushion starfish. Despite its name it has a hard chalky skeleton

layer of hot water floating on a layer of colder water. Since there is very little circulation between layers, there is little diffusion of dissolved minerals and those in the top layer become exhausted by the organisms living there without being readily replaced. When these organisms die, many of them fall to the deeper water where they decompose, so that their mineral remains are also trapped. In the oceans, and in many of the other seas of the world, the thermocline is broken up by stormy weather, particularly in the autumn, but the Red Sea is normally calm for most of the year. Thus, although its depth may vary, the thermocline persists. One of the effects of this is to limit the growth of phytoplankton (floating microscopic plants) upon which the zooplankton (floating microscopic animals) depend for food. Because the plankton in the Red Sea is somewhat sparse the community of pelagic fishes, such as flying fish and herring-like fish is also reduced.

The fringing reefs along the Red Sea coasts, especially on the African side, generally enclose a sheltered lagoon. They present a continuous wall of coral limestone which to some extent protects the shore from the forces of the sea. They are, however, broken at periodic points where water courses enter the sea. These water courses are normally dry, but may be raging torrents for a few days in the year after rain has fallen. Why the fringing reef should be broken by the influence of the water courses is not exactly clear, but it is probably due either to the fact that the coral polyps cannot withstand the occasional smothering by silt borne down by the water, or to the fact that they are killed off by the sudden fall in salinity. These breaks in the reef, together with the region where the water course passes into the lagoon are known as 'mersas' by the local inhabitants. Such mersas provide excellent and convenient harbours for small craft and are used as such to a considerable extent by local fishermen and traders.

It was in a mersa harbour that, in August, 1971, the author joined a large motor launch carrying the scientific equipment for a research expedition to one of the more distant off-shore reefs. The launch, with a group of 12 people, was not due to leave until the following morning so the night was spent sleeping on the upper deck and cabin roof. Since the trip was taking place during August the weather was extremely hot and the sea was very calm. The water in the

Right: A narrow channel links this lagoon with the open sea and may form the only access for boats. Strong currents will flow through it as the tide ebbs and flows

lagoon was almost motionless, disturbed only occasionally by the stirring of a large fish.

The following morning the launch set out for the diving point. As it approached the landward side of the outer reefs the sea became a progressively paler blue, indicating the upward slope of the sandy sea-bed. This gradual upward slope contrasts markedly with the precipitous drop of the seaward edge of the outer reefs, and also the sharp drop at the outer edges of the fringing reefs.

Any dive must be preceded by the correct planning and preparation. This is especially true when scientific observations are to be made or samples taken. Apart from checking out the equipment and ensuring that it is correctly assembled for use, each member of the team must be briefed as to his specific role and responsibilities. On this occasion eight men were to dive as four pairs to collect specimens of the Crown-of-thorns starfish *Acanthaster planci,* for use in Y-mazes on the sea-bed. This was part of a series of experiments meant to determine which corals the starfish preferred to attack. The mazes had been laid down previously at a point near the launch's anchorage. During the dive each pair was to survey the colonies in a portion of the main reef for the characteristic white predation scars, locate and collect the starfish themselves, and transport them carefully back to the experimental cages. In addition, the divers were to note which coral colonies had been attacked, what their positions were on the reef, and whether or not the individual starfish showed any signs of attack from other organisms. White formica boards with lead pencils served as effective underwater note-pads for this purpose.

The divers were transported to their sectors of

Below: Coral reefs offer exciting locations for underwater fish enthusiasts, whether they dive to spear fish or to take photographs

the outer face of the reef in inflatable dinghies which had previously been towed behind the launch. In them were carried all the necessary pieces of equipment, including plastic containers for transporting starfish. At predetermined spots the dinghies anchored and each pair of divers, now equipped with their scuba gear and collecting apparatus, rolled into the sea over the sides of the rubber inflatable.

Since the divers had prepared correctly for the dive, once they were overboard the weight of their air tanks and of the lead on their weight belts created neutral buoyancy. Diving in the Red Sea is always a great pleasure for at least three principal reasons: the beauty of the marine life; the fact that the sea is so warm in summer that a diver can stay in it almost all day long; and the very good visibility, especially near the surface where the illumination is good.

Two members of the team found themselves about 15 m (50 feet) away from the reef's edge. By swimming almost vertically to a depth of 10 m on their depth gauges, (about 33 feet), they had reached a point on the reef slope where it flattened out in the form of a terrace. Their task now involved swimming along this terrace searching for the starfish predation scars among the coral growths.

Looking upwards they could see the clear outline of the inflatable dinghy. Turning to look up the slope of the reef towards the reef crest they could see a variety of coral colonies both on the inner edge of the terrace, not far from where they were kneeling and then more rising on the steeper face of the reef as it rose sharply towards the surface. Around many of these colonies were shoals of brilliantly coloured small fish, shimmering with constant movement.

Below: Colonies of various corals in the lagoon of Moorea in the Society Islands, growing in the strong light just below the surface

Marina Cedri/Rives

Above: A giant clam, Tridacna *species, with fleshy mantle tissue protruding between the open valves. Such clams often become embedded as corals grow around them*

Above left: Feather-stars are among the most attractive creatures found on coral reefs. The mouth is in the middle of the animal and food particles are swept along the arms to it

Left: A brightly coloured sea-slug

The coral growths on the terrace consisted mainly of small to medium-sized colonies, from about 25 cm (10 inches) to 100 cm (39 inches) across at their widest points. Some of these were tabular growths of *Acropora,* and on one of the first of these which the two divers approached approximately two-thirds of the polyps had been killed and eaten by the Crown-of-thorns. Resting close by in a crevice in the coral limestone were two specimens of *Acanthaster planci* each measuring about 15 cm (6 inches) in diameter. It was apparent that these two starfish had been feeding on the same *Acropora* colony for about three or four nights. The divers knew this because close inspection of the whitened part of the skeleton revealed that the inner part was faintly tinged with green, whereas the outer part was pure white. This green tinge was the result of the colonization of the dead coral skeleton by minute single-celled plants (algae), which happens almost as soon as the digestive action of the starfish has finished cleaning the skeleton. After a few days, when the starfish have completed their attack, the whole colony would have become a dingy green due to the increasing growth of algae. One of the divers, using a long steel rod, touched the parts of the starfish which were innermost in the crevice to encourage them to move into the open where they could be picked up without damage to either animal or diver. Once they were in the open, they were quickly checked for missing rays or other unusual characteristics and were then carefully placed in string bags. Living on the outside surface of one were a pair of minute shrimps, *Periclimenes soror*. The relationship between these animals and the starfish is not fully understood, but it is likely that both the shrimps and the starfish gain some advantage. The shrimps were coloured in such a way as to make them inconspicuous against the reddish-brown background of the starfish, and thus presumably to reduce their chances of being snapped up by passing fish.

The divers then continued along the terrace, encountering several more intact coral growths of *Acropora* which were surrounded by swarms of small green damsel fish (*Chromis* species) and little black and white striped 'Humbug' fish, *Dascellus aruanus*. At one point along the terrace a great quantity of *Acropora* had fallen down from higher on the reef face and formed a sort of 'landslide' of broken coral fragments. Large colonies of coral sometimes become unstable if they grow unevenly, and are more susceptible to being broken or overturned in rough weather. This coral had been dead for some time, and all the broken branches were encrusted by a variety of plants and animals. Apart from the brown-green filamentous algae, it was ornamented by colonies of hydroids, sea-mats, small brown and grey sponges and tube-worms. One of the divers picked up a handful of dead coral and exposed a previously hidden community of black brittle-stars, feather-stars, sea-urchins and some turban shells. All these animals live among the coral remains, and emerge at night either to feed on the small encrusting organisms growing on the dead *Acropora* itself, or to strain plankton drifting past in the water currents.

Just beyond the 'landslide' of *Acropora* the divers noticed a group of three small, rounded, unbranched coral colonies, two *Goniastrea* species and one *Favia* species. They were all dead and whitened, obviously prey to another Crown-of-thorns. Nearby, in a small crevice, the entrance to which was guarded by a small brilliantly red sponge, was another starfish which was placed in a string bag when the usual details had been noted. One arm of this specimen was missing but showed signs of regeneration as a small stump jutting out between the remaining ones. Not far from this point, on a small outcrop of limestone, was a large colony of soft coral. From a distance it somewhat resembled an animated cauliflower

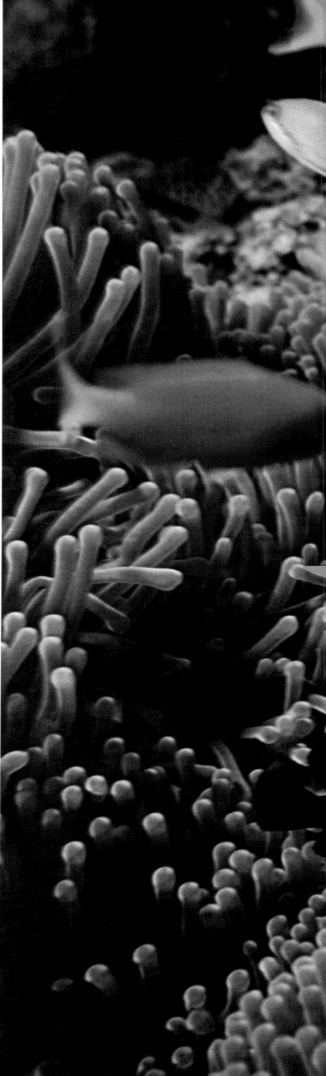

Right: Clown fish are easily recognized by their orange-yellow bodies which are marked with bluish-white vertical stripes. They usually take refuge within the stinging tentacles of a sea-anemone

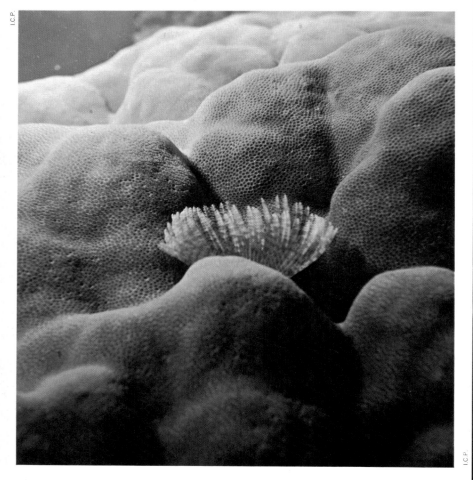

Below: A fan-worm in a colony of one of the true corals, a species of Porites

Left: Spectacular cascading colony of soft corals

Below left: Tropical sponges such as this one, with its decidedly unusual appearance, are not hard to find among the coral formations of the Red Sea

Below: A cluster of flask-shaped sponges photographed in the Red Sea. Water enters through the small holes and is expelled through the large hole

from which the leaves had been stripped, leaving the greenish white interior. Each polyp was repeatedly opening and closing its eight branched tentacles and drawing them across its mouth.

The divers had now reached a position on the reef where its outer edge changed direction, and the terrace abruptly ended. Instead massive pillars of limestone rose abruptly upwards from a depth of about 60 m (200 feet) and the lack of delicate branching coral growths suggested that this position was more exposed to wave action, presumably on account of its facing North-east instead of East. By following the line of limestone butresses the divers were in fact passing through a channel in the reef which cut it into two sections and led from the seaward side into the sheltered water beyond. All the coral colonies that occurred on these steep limestone faces were of small, stout and rugged varieties. They included examples of *Montipora*, *Pocillopora* and small brain corals. Between most of these colonies were considerable expanses of exposed rock covered only with thin growths of algae. These areas, where wave action in rough weather is too strong for branching fragile coral colonies, have their own characteristic inhabitants. Boring a burrow into the limestone were small black sea-urchins, *Echinostrephus molaris*, which collect drifting particles of algae broken from the rocks by waves and currents. When a particle collides with the spines they converge and hold it until it can be passed around the edge of the urchin by means of its tube feet and down into the burrow. Touching the long top spines of one of these animals by hand usually causes it to retreat into the burrow. It appears to be quite efficient at collecting and storing drifting objects and one charming report records that specimens were discovered in the Indian Ocean storing grains of rice, presumably thrown overboard from a ship anchored nearby, in their burrows. There are many other animals which invade or encrust the limestone in areas like this, and the divers noticed tubes set in the rock with openings about 2 cm (0·8 inches) in diameter and apparently stopped by a plug. Tubes like this are occupied by a form of highly modified sea-snail belonging to the genus *Vermetus*. The sea-snail filters plankton from the seawater passing over its tube mouth, and does not feed as a grazer like many of its relatives. Also protruding from the limestone were the smaller tubes of a number of tube-worms, from which extend a coloured crown of tentacles which act as filter feeding organs and also assist in respiration.

When the divers had reached the far side of the

reef they found that in the more sheltered water behind it the sea-bed sloped gently down to a sandy bottom at a depth of about 40 m (130 feet). A profusion of small coral colonies and soft corals grew on this slope among which were a number of fine tabular growths of *Acropora* reaching nearly 200 cm (78 inches) across in some cases. Several of these showed signs of predation by *Acanthaster*, and under one of them the divers spotted a group of ten of the starfish sheltering from the sunlight.

Close by was another small crevice, and from its opening protruded a number of long, needle-sharp black spines indicating that the niche was occupied by one or two large black spiny sea-urchins, *Diadema setosum*. At night time, these sea-urchins leave their shelter and browse over the reef, attacking algae, sedentary animals and even the coral polyps themselves. *Diadema* is not as well protected from its enemies as its spiny armament would suggest, and urchins on the reef by day are often attacked by passing fish such as the trigger fish, *Pseudobalistes flavimarginatus*. These fish are able to take the sea-urchin by surprise, blowing it over with a jet of sea-water puffed from their mouths. When the urchin is upside-down, its under surface, where the spines are shorter, is vulnerable to attack by the fish's jaws and teeth. The so-called test, or shell, is then cracked open and the fish extracts the roe for food.

A related tropical Atlantic species *Diadema antillarum*, is attacked very successfully by the helmet-shell of the genus *Cassis*. The helmet-shell, a large gastropod sea-snail, will approach an urchin having detected its presence, presumably, by smell. When the urchin detects any dark object or shadow approaching it, all the spines converge towards the direction of the approach. When within reach the helmet-shell rears up on its foot and smashes its shell amongst the spines. Having got this close it can spray the outside of the sea-urchin with its saliva which contains a drug. The drug acts to immobilize the spines, presumably by blocking their nervous co-ordination system which lies just below their surface. While the urchin is thus narcotized, the helmet-shell cuts a hole in the test with its

Far left: The heterogeneous community which inhabits the waters of the Red Sea; a great variety of corals dominate the scene

Top left: The nocturnal Slate-pencil sea-urchin. Its club-shaped spines have been used on some islands as pencils for writing on slates

Left: A brightly coloured jellyfish, an animal distantly related to coral polyps

ribbon-like tooth apparatus or radula, and sucks the roe from the cavity.

The divers prepared to collect the specimens of *Acanthaster* they had seen underneath the tabular *Acropora*. One of them was about to rest his elbow on a small rocky outcrop, when the object he had thought was a rock darted away, travelling about 20 m (65 feet) only to settle again and melt into the seascape. It was in fact a stone fish, a member of the family Scorpaenidae, measuring about 50 cm (20 inches) in length. Small stone fish are common on the floors of lagoons where their poisonous spines, which can inject powerful venom, form a great threat to bare-footed bathers. This specimen was quite large and would have certainly injured the diver had he touched its poison spines. Its camouflage is usually excellent, so that the stone fish can just lie in wait for some small animal to swim within reach of its powerful jaws.

Having successfully collected all ten Crown-of-thorns the divers swam slowly towards the surface over the inside edge of the reef to call their boat. As they were waiting at the surface they could hear the sound of pistol shrimps snapping their claws among the rich growths of *Acropora* and *Pocillopora*. There are many species of pistol shrimp and more than one genus. They generally have one claw or nipper far better developed than the other. By slowly contracting the muscle of the nipper, but not allowing it to shut they can build up great tension in the joint. Then, when suitable prey approaches they can release the claw which snaps shut with an audible click. Although the shrimps may be only a few centimetres long, this click is often loud enough to temporarily stun small passing animals which are then siezed and eaten. Apart from the pistol shrimps these colonies had a number of other crustaceans associated with them, including a variety of brightly coloured crabs and other shrimps and prawns. Also there were black brittlestars and a variety of damsel fish.

In a large fissure, just to one side of the *Acropora* colonies, was a group of six giant sea-anemones, species of *Stoicactis*. The anemones were waving their poisonous tentacles slowly back and forth as the small waves washed over them, and among the tentacles swam specimens of a well-known clown fish, *Amphiprion bicinctus*, and the Domino damsel fish, *Dascellus trimaculatus*. Although several species of clown fish are associated with such anemones all over the tropical Indo-west Pacific, it appears that only in the Red Sea has the Domino damsel also taken on this behaviour. In other parts of the Indo-west Pacific they may live in the vicinity but not take refuge among its tentacles. What the mechanism is which prevents these fish from being stung to death by the anemone is not fully understood, but recent scientific work on the two partners suggests that the large tropical anemones may not be as lethal to small fish as some of the smaller species occurring in both tropical and temperate waters. A clown fish which is confronted by a new anemone appears to have to spend some time getting to know it. It does this by swimming round the anemone and rubbing some of the anemone's slime on to its back, fins and sides. Possibly this mucus which is secreted by the anemone confers protection on the fish by reducing the firing of the anemone's stinging cells. Certainly, the act of taking a fish from one anemone and thrusting it into another's tentacles results in damage or death for the fish. The question must arise as to what advantages the

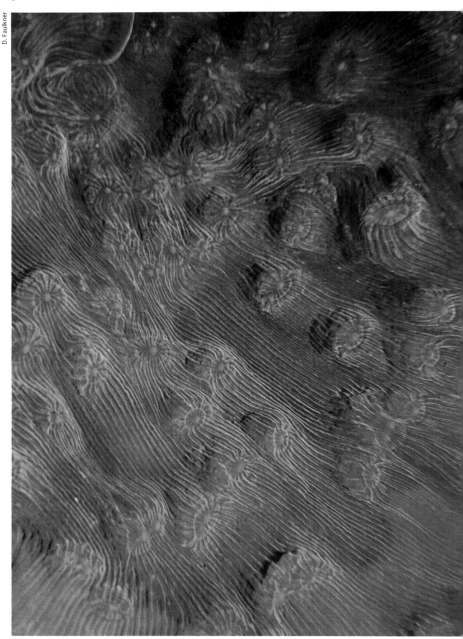

Below: Flourescent effect on flat growths of the true coral Mycedium

association holds for the partners. For the fish it is probably protection from enemies, but for the anemone it would appear to be grooming. The clown fish, particularly, may spend some time passing the various tentacles through their mouths and picking away injured parts or parasites. They may also bring scraps of food back to the anemone as a result of their foraging.

The boatman had now brought the inflatable to the divers and passed them the plastic tanks which they submerged briefly so that the string bags of starfish could be emptied into them without the animals being exposed to the air. The tanks were then lifted back into the boat and covered with wet sacks to shield the animals from the sun, and to keep them reasonably cool by evaporation. The divers then heaved themselves into the inflatable and removed their equipment during the run back to the launch. After debriefing and a discussion with the other members of the team it was decided that a night dive would be planned during which the captured animals would be placed in the previously prepared underwater Y-mazes, and observations made on any other specimens of *Acanthaster planci* which might be feeding in the area. During the intervening period all the collected starfish were placed in a large aerated tank of sea-water on board the launch.

It was quite dark when the four divers who were to participate in the night dive climbed back into an inflatable. During the afternoon another member of the team had fixed a small torch to the buoy marking the Y-mazes so that by night it was a relatively easy matter to find it. Each diver carried an underwater torch and had been assigned two string bags of starfish which at the moment were still immersed in sea-water in the

Below: Part of a coral outcrop standing on the sandy bed of the Red Sea. A mixed shoal of Anthias *and* Chromis *are feeding near a colony of* Acropora

plastic tanks. When the inflatable reached the buoy it was tied to it and the engine was cut off. The sudden cessation of noise revealed the quietness of the sea at night, the only regular sound being the breaking of small waves against the reef edge and the side of the boat. The divers quickly put on their scuba gear and slid into the water. Then they were passed their torches and finally the bags of starfish. It was a relatively easy task to follow the buoy's cable down to a depth of about 10 m (33 feet) where it was tied to one of the mazes. Shining the torch ahead revealed a sort of fog in the water, caused not by sediment but by zooplankton which at night rises to the surface of the sea. The zooplankton quickly made itself felt too. None of the divers were wearing wet suits because of the warmth of the water, so they were constantly stung by numerous small planktonic jellyfish which could just be made out in the light of the torches as minute white flecks. Another striking feature of the plankton was the phosphorescence emitted by microscopic crustaceans whenever the water was disturbed by a flipper or an arm. At the bottom of the cable the wire-netting sides of the Y-mazes emerged from the gloom. Although they had been in position for only a week or so, there were already signs of algal growth on the wire mesh, making it look coarse and shaggy. The string bags of starfish were emptied into their respective wire enclosures, and once this operation had been completed by each member of the team, the divers swam towards the reef using their compasses for orientation.

The first pair of divers arrived at the reef base near a fall of dead *Acropora* similar to the one which they had encountered earlier during the day. By night, however, it seemed to have become alive. Small brownish-red squirrel fish with huge eyes were swimming with a jerky motion over the branches, scattering away from the direct light of the torches, while climbing over the dead pieces of coral were a number of echinoderms including sea-urchins and feather-stars. The feather-stars were filtering food from the sea-water with their extended, branched arms.

On a patch of sand, beside the base of the *Acropora* slide, one of the divers picked up a medium-sized starfish with seven arms or rays. This was a specimen of *Luidia savignyi* which belongs to a group of starfish renowned for their fragility. As though to demonstrate this fact, the arm it was being held by suddenly broke from the rest of the body which fell to the sea-bed and moved away.

The divers then swam further along the edge of the reef in search of feeding *Acanthaster planci*.

They noticed seven specimens of the sea-urchin *Diadema setosum*, with their very long spines waving back and forth in the light of the torches. They were grazing on a patch of exposed coral limestone overgrown with algae and encrusted by animals. These sea-urchins, which are familiar inhabitants of most reefs, are responsible for keeping down the population of encrusting organisms. If they are excluded from a patch of reef, by covering it with a small wire cage for instance, a prolific crop of algae and small sessile animals will quickly develop.

After searching the reef for feeding *Acanthaster* and attaching marker buoys to the sites they had found so that they could be studied the following day, the divers explored the slope of the reef near the surface. Just below the surface, moving slowly between some large colonies of *Porites* were two cumbersome looking Slate-

Above: It is quite common to see small sharks in shallow water among the reefs. A nurse shark like the one towards the top of this picture may reach 4 m (13 feet) in length

Above: Night diving on the Great Barrier Reef. There is a large sea-fan on the right displaying the characteristic delicate tracery

pencil sea-urchins, *Heterocentrotus mamillatus*. These animals, with their brown tests and thick red spines, are adapted to life in the surf zone, and in the daytime they wedge themselves into crevices of the reef using their powerful spines as anchors. At night they emerge to graze in a similar way to *Diadema setosum*.

Just beside the Slate-pencil sea-urchins was another sea-urchin which none of the divers had seen before, and which responded directly to the light of the torch by hastening away. In order to get a better look one of the divers who was wearing gloves made a grab for it, but he had hardly touched it before he was writhing in agony at the surface. Fortunately it was possible for him to stagger on to the top of the reef and to get quickly into the inflatable. The urchin that he had touched was *Echinothrix diadema*, one of the most unpleasant animals on the Red Sea reefs. Although it is not as poisonous as some sea-urchins, it is capable of inflicting severe stings by means of its hollow secondary spines. The larger primary spines are blunt, although also hollow, but the secondary ones are needle sharp and equipped with fine barbs so that they are very difficult to remove from the wound. In this instance seven such spines had penetrated the rubber glove and passed under the nail of the diver's thumb and forefinger, staining the affected areas purple-brown and causing considerable pain. Whether or not the fluid in the hollow cavity of the spines is itself the poison is not yet established, and virtually nothing is known about its pharmacological action.

As a result of this incident the rest of the night dive was abandoned and the divers returned to the launch, where they could prepare for the next day's work and get some sleep.

The fish of the coral reefs

As the descriptions in this book have made clear, coral reefs offer an enormous variety of habitats to marine life other than the coral itself. Almost every available position on a reef is colonized by a plant or an invertebrate animal which may be a filter feeder, a herbivore or a carnivore. Fish, too, appear to exploit almost all the ways of life made available by a coral environment.

There are many standpoints from which a review of coral reef fishes and their behaviour could be undertaken. Some fish are territorial, inhabiting a particular area in the region of an identifiable rock or crevice. Many of the smaller fish are not limited to any particular point, but nevertheless associate with only one type of coral colony or rock formation and can move from one to another when necessary. Other fish are even less restricted in their movements, spending their lives cruising around the reefs and through the lagoons in search of food, and they may live singly, in small groups or in large shoals. Truly open water species may pay visits to the reefs, swimming just within visual range of the reef edge and making occassional foraging sorties into the reef community.

Other aspects of fish biology vary apart from those associated with habitat. Reproduction provides a whole spectrum of varieties and evolutionary levels. Firstly, unlike many temperate water fishes, coral reef fishes are often highly patterned and coloured. Such patterns and colours may be associated with the process of courtship and breeding, or with the establishment of territory. In an area where many closely related species are to be found it is essential for members of one to be able to distinguish between their own kin and those of near or more distant relatives. Males and females are often identified by their different colorations, and both colour and pattern may change when the breeding season starts. Secondly, although in most species there are separate males and females, a few groups like the wrasses are hermaphroditic. In these cases the fish start life as one sex and remain in that sex for several seasons after they have reached maturity, then changing to the other.

There is one way in which the community of fish on a coral reef can be divided into two populations: those which forage for food by day (diurnal), and those which do so by night (nocturnal). A diver carrying out a survey on a reef in the daytime would find, when the same area is revisited at night, that a completely different set of fish are active. A very close and detailed survey of a reef in the daytime would reveal a number of the nocturnal species hiding in crevices and under coral outcrops; by night it is generally rather easier to find diurnal fish sleeping among branching corals and in caves.

One result of this division into diurnal and nocturnal activity has been the evolution of an interesting phenomenon known as resource sharing within the community. An example is given by the avilability of shelter sites. During the day a crevice in the reef may be occupied by a sheltering nocturnal fish. At dusk this crevice is vacated, and the space can then be used by a diurnal fish which needs somewhere to hide at night. Such sharing activity may be extended to other forms of behaviour too, such as areas for hunting and feeding.

Left: Just as a forest provides shelter for land animals, so huge branching Acropora *colonies can play host to fish and other marine organisms, providing both food and shelter*

Left: Two rabbit fish in the Red Sea. These fish are so-called because of their rabbit-like mouths

Below left: Juvenile butterfly fish. Young members of this genus, Chaetodon, have a distinctive dark band across their eyes and pointed mouths. The adults are usually differently marked

Below: A Clown fish lives as a symbiont with sea-anemones. Once this relationship is established, the fish appears to be immune to the stinging tentacles

Around dawn and dusk it should be possible to observe the so-called 'change-over period' when diurnal fish are settling down for the night and nocturnal ones are emerging for their activity period. Some species, such as certain puffer fish, undergo marked colour changes when they get ready to spend the night in the seclusion of coral colonies. Others, like the parrot fish, undergo even more profound behavioural changes. The parrot fish secretes a mucous envelope or 'night-shirt' as a result of the activity of special mucous glands. A new one is 'put on' every night, and it has been suggested that this is a means of containing any scent that might be released by the fish while it sleeps.

A further interesting point about reef fish that has recently attracted the attention of researchers relates to the sounds made by the fish. Many of these are audible to the diver, and others may be detected with the aid of special equipment. Scientists have been able to correlate these sounds, grunts, squeaks and other noises with various aspects of the life of the fish concerned like territorial behaviour, aggression, feeding or escape. The species of fish making the noise, as well as its mood, may be detected, and these sounds are presumably of great significance to other members of the reef community.

These, then, are a few of the many variations in behaviour from which a survey of the coral reef fishes could be conducted. Ultimately, however, the most convenient point of view from which to look at the fish community is that of feeding habits, and this is the approach that has been adopted in the remainder of this chapter. It must be remembered, however, that very few examples can be discussed here – the intention is simply to give a brief picture of the fascinating range of behaviour among coral reef fishes.

Although there is no shortage of small filamentous algal growths covering large areas of rock face with so-called 'algal lawn', this does not mean that many reef fish rely directly on algae as a source of food. Indeed, the herbivorous fishes encountered on coral reefs may be regarded as being more advanced on the evolutionary scale than the carnivorous ones. Carnivorous fishes rely on other animals, mainly invertebrates, to convert plant protein into animal protein for them. Very few groups of fish are composed of only herbivores; most comprise a mixture of both herbivores and carnivores. Several groups like this are well known on coral reefs. One such is the family Acanthuridae, the surgeon fishes, which are prominant residents in many parts of the tropics. They are so called because of their ability

to extend a small, very sharp lancet-like protuberance on either side of their tail stalk. Contact with one of these protuberances, which are developed from modified scales, may lead to a most unpleasant cut, so they are well named after the surgeon's knife. Surgeon fish are ovoid in outline, often they have delicate trailing tail fins and small mouths. They may be very colourful so that the shoals which these fish form can provide quite an underwater spectacle. It appears that they are able to change their coloration according to the time of day and their mood. Particular species also show quite well-defined racial markings in that their coloration varies from one part of the tropics to another. Many of the species of surgeon fish are able to breed all the year round. The females lay many thousands of eggs which float in the sea after they have been fertilized. This inbuilt buoyancy is due to a minute oil droplet inside the egg. Young fish usually hatch out after a period of around 26 hours and spend a brief time using up their yolk supply before they start to feed on other planktonic organisms.

Many adult surgeon fish feed by rasping on rocks to pick up algae growing there as well as encrusting invertebrates. Some have become specialized to pick over softer substrata such as shell-gravel and sand. Particles of the substratum are ingested along with any algal matter growing there. Some of these fish are able to bite off portions of the more leafy algal growths. A few surgeon fish are planktivorous, that is they feed on planktonic organisms.

Where currents of water well up at the edges of the reef, they bring food and oxygen towards the surface. The degree to which this upward movement of water occurs depends on a variety of factors, including the effects of prevailing currents, tides, and diurnal temperature fluctuations in the sea. A variety of small fish forage in these water columns for small items of planktonic food like invertebrate larvae, crustaceans and other tiny animals. One group of fish which seems particularly to depend on this type of food being brought to the surface waters is the damsel fish (family Pomacentridae). These generally small fish always shelter at night 'roosting' in groups like birds, among the branching corals. After dawn the fish leave their coral hiding places and rise higher into the water as the day passes and the column increases. When danger threatens, some damsel fish such as species of *Chromis*, form tight shoals which head quickly for the cover of nearby coral colonies. This behaviour reduces the chances of successful attack by larger predatory fish as it eliminates stragglers. At night the position of *Chromis* species on the reefs is often taken over by cardinal fish such as species of *Apogon*. These nocturnal fish come out to hunt for small swimming crustaceans.

The damsel fish are among the smallest fish to feed on plankton; they seldom exceed 6 cm (2·4 inches) in length. Other very much larger fish such as the Giant manta ray, an occasional visitor to the reefs, are also planktivorous. This vast fish can attain a maximum width of 6 m (20 feet) and may be almost as long. On either side of its slit-like mouth is a flap-like appendage which acts as a

Above left: Brightly coloured damsel fish

Above: Porcupine fish (Diodon hystrix) *in the San Blas Islands off Panama. These fish are skinned, inflated and dried for sale as souvenirs by the islanders*

guide to the inflowing water. The water enters the mouth and leaves via the gills due to the fish's forward progress. As it does so the water is filtered for plankton by special structures in the gills. Periodically the captured particles are flushed off the filters and swallowed. The Whale shark, a relative of the rays, is probably the largest fish in the world, reaching 15 m (50 feet) in length, and it too is a filter feeder. It mostly inhabits tropical waters, feeding in much the same way as the Giant manta ray.

Although there may be several representatives of the cartilaginous or shark-like fish in the vicinity of most reefs, it is the bony fishes principally which have diversified to occupy so many different niches in the reef habitat. Among them is one group especially typical of coral reefs, the parrot fish (family Scarridae), which have for a long time been regarded as important predators of coral. The teeth in their upper jaws are fused, giving rise to a structure resembling a parrot's beak. This feature, together with their often gaudy colouring, makes these fish attractive and conspicuous as they swim lethargically about amongst the coral colonies, 'flapping' their pectoral fins as a means of locomotion. A few species appear to prey largely on massive coral colonies like *Porites* where their characteristic teeth marks can often be found. As they bite into the coral the sound of their teeth breaking into the calcium carbonate can be quite audible underwater. For each mouthful that they take in they must consume a great amount of mineral matter and very little coral tissue. The majority of parrot fish species, however, do not actually attack live coral colonies, but choose those which have been dead for some time and whose outer surfaces are covered with growths of algae. This suggests that parrot fish are principally herbivores and not carnivores as had been previously supposed. Since they are relatively abundant they must be significant in keeping down algal growths.

The family Scarridae also includes a few types of larger fish like the so-called Donkey fish,

Below: The striped Humbug fish maintains a constant presence near colonies of Acropora, *sheltering at night in the branches. Also seen amongst these branches are less distinctive damsel fish*

Bolbometropon muricatus, which occurs in the Red Sea. This large fish may reach 2 m (6 feet) in length and it feeds on coral colonies. In order to break the colonies up into manageable sizes it charges at strands of coral and knocks them down with the enormous hump on its head. It is then able to deal with the fragments using its powerful teeth.

Not only does the sea around coral reefs offer various forms of planktonic food for fish, the choice of invertebrate prey on the sea-bed is just as great and has proved attractive to many predatory fish. Surrounding the base of many reefs is a wide expanse of coral sand which slopes only gently in comparison with the steep-sided drop of the reef itself. This sand is formed from ground coral pieces which have originated from the reef, and it generally contains a wealth of organic matter, deposited through the death and decay of fish and other swimming organisms which have fallen on to it, as well as by the death of animals which burrow into it. A number of fish display adaptations which enable them to hunt for invertebrates living on or in the sand. Some of the attractive small rays which swim around the reefs spend a great deal of time hunting on the sand, and if any animals are detected under it they can be uncovered by the ray and crushed between its modified teeth which are arranged on its jaws like tiles, and are ideally suited to crushing shelled organisms. Among the bony fish, some like the goat fish (family Mullidae) use their special sense organs to detect the presence of food below the surface of the sand. These organs take the form of small barbels (fleshy filaments) which protrude forward from the underside of the lower jaw. Some of these mullet-like fish also supplement their diet by taking planktonic food. Most of them hunt during the day time.

Apart from the invertebrates which burrow in the sand there are those which dwell among the corals themselves, living in crevices from which they can emerge occasionally to feed. Other invertebrates are actually attached to the rocks themselves and depend on currents to bring them their supplies of food and oxygen. The range of shapes and styles of invertebrate reef life is immense, and some of these animals, like the sea-urchins, have well-developed defence mechanisms, so that fishes which prey successfully upon them have to be quite specialized. One family which shows remarkable specializations to this effect is that which includes the angel fish and butterfly fish (family Chaetodontidae). These are very attractive reef fishes which can often be distinguished easily from other types by virtue of their triangular outlines and long dorsal fin rays. A pointed snout is another general feature. These fishes have developed a variety of life-styles and habitats; one reef may contain many different species, all of which can be distinguished from each other in details of shape and colour. Many of them are specialized feeders restricted to only one sort of invertebrate prey. An example is an angel fish, *Holacanthus arcuatus,* which according to recent reports feeds almost exclusively on sponges. Any other material found in its stomach appears to have been incidentally eaten— probably having grown on the outer surface of the sponges. Other species of angel fish are not so limited; some feed on the algae and organic detritus covering the sea-bed.

A characteristic feature of the butterfly fish is their somewhat pointed and extended jaws. In many species these protrude for only 1 cm (0·4

Above and top: The Sea-scorpion, like the Stone fish, has an astonishing ability to change its coloration so that it can camouflage itself on any kind of sea-bed. These fish are often dangerous because of their poisonous spines

Above: A group of Domino damsel fish keep station by a cluster of anemones in the Moorea barrier reef in the Society Islands

inches) or less, but in some cases they are much longer, reaching up to 3 cm (1·2 inches) or more, as in species like *Forcipiger longirostris*. Jaws of this length appear to be a modification for searching in crevices of the reef and among coral branches to take up whole specimens of small prey species like shrimps. The backwardly pointing teeth at the tips of the jaws, which effectively act like forceps, presumably assist in gripping such animals. A close relative, *Forcipiger flavissimus*, has a somewhat shorter snout and a greater area of gripping teeth. This species, it seems, attacks larger prey and tears pieces from them instead. The animals that it attacks are usually bottom dwellers. Butterfly fish with the shortest snouts are generally not so specialized. They prey more on small planktonic and benthic (living on the sea-bed) animals taking small bites from them. With benthic prey such as soft corals this is quite an easy process of feeding. The impression of diversity of butterfly fish that is conveyed to the diver or fisherman when he looks around a reef is usually somewhat accentuated by the fact that juvenile specimens may appear differently patterned or marked from the adults of the same species. Further, juveniles and adults may occupy different habitats which is advantageous because the two do not then compete for food or space. Both as juveniles and as adults these attractive fish tend to swim around in pairs or small groups.

The wrasses (family Labridae) comprise a very large and important family of marine fishes which inhabit shallow water and rocky coasts all round the world. In the tropics they have diversified to occupy a great variety of habitats and to follow many different life-styles. Many wrasses are fairly deep bodied and beautifully patterned fish, while some are drab and others are thin and slender. Generally they are predacious, feeding on other fish and invertebrates. Those which take shelled invertebrates as food, like the Hump-headed wrasse, *Cheilinus undulatus,* have specially modified pharyngeal teeth at the entrance to their gullets. These are massive plate-like structures which crush shells and skeletons, and enable the fish to consume sea-urchins, sea-snails, and other hard objects. The Hump-headed wrasse is very large, growing to a length of 100 cm (39 inches) or more and reaching a weight of 50 kg (110 lbs). By virtue of its relatively plain patterning and dull coloration it is able to swim slowly round the reef and approach and attack unsuspecting invertebrates. Examination of gut contents indicates that this species may feed on invertebrates like Slate-pencil sea-urchins, *Heterocentrotus mammillatus,* which are active on the reef at night time.

Small to moderate-sized wrasses abound at almost every point around the reef. One example, *Thalassoma duperrey,* feeds in a variety of ways, foraging for plankton in the water column, searching crevices, feeding on invertebrates in sand, and following herbivorous fish to eat any small animals that their grazing activities have uncovered. These fish and other similar species are always on the look out for food and will quickly approach a diver if he has made a new supply available to them.

Among the smaller wrasses are the remarkable cleaner wrasse, of which a considerable number of species have been identified throughout the tropical world. One of the best known is *Labroides dimidiatus* which has become specialized to remove ectoparasites from the bodies of other fish. They position themselves at particular points along the reef which become established 'cleaning stations', and which may be visited by other fish wanting to be cleaned of parasites. *Labroides dimidiatus* swims about in a characteristic undulating fashion, and this behaviour, in conjunction with its blue and white striped coloration, makes it conspicuous. It has a fork-like cleft in the lower jaw which it can use to prize off parasitic animals rather as one would use a nail extractor to get old nails out of pieces of timber. When confronted by another fish, which indicates by its posture that it needs to be cleaned, the cleaner swims back and forth along its 'patient' searching for growths and parasites. It will even enter the posturing fish's mouth to remove foreign bodies and it will also clean between the gills.

One of the most fascinating instances of form and behaviour on the reef is provided by another fish, the Sabre-toothed blenny, *Aspidonotus taeniatus.* This fish bears more than a superficial resemblance to the cleaner wrasse, having a similar shaped body and similar coloration and patterning. The blenny can be distinguished, however, by the position of its mouth which is on the underside of the head, rather than at the tip of the snout as it is in the wrasses. A further difference lies in the way in which it swims, using its tail much more than do the cleaner wrasses who tend to manoeuvre with their pectoral fins. The Sabre-toothed blenny also occupies the stations that are set up by the cleaners, and fish needing cleaning attention will posture themselves in front of it in just the same way. Rather than cleaning them, however, the blenny swims over their bodies and bites pieces out of their fins. This fascinating example of mimicry thus provides the fish with its feeding opportunities.

Right: A large reef anemone, almost closed. Some small clown fish (genus Amphiprion) are associated with it. These fish probably groom the anemone and drop small pieces of food in its tentacles in return for the shelter it provides

Below: A carnivorous ray. This is a cartilaginous fish like the shark and dogfish. Some species carry an organ capable of inflicting quite severe electric shocks, probably used for stunning prey as well as for defence. Other types have stings on the tail

82

Right: Moray eels are commonly found in cracks and crannies in coral outcrops. They stay hidden by day and emerge to hunt by night

Left: A Scorpion fish or Lion fish (Pterois radians) in the Red Sea. It is one of the most beautiful and distinctive fish in the reef, and also one of the most poisonous

Below left: Example of Holocentrus rubus, photographed among corals. The large eye is a modification for nocturnal life

The puffer fish (families Tetraodontidae, Canthigasteridae and Diodontidae) have many representatives which prey on invertebrates. Often they have a few powerful teeth at the front of the jaws which enable them to deal with spiny or shelly prey. They are known as puffer fish because of their fascinating habit of inflating their bodies to greatly increase their size. In this way they may threaten or frighten potential enemies. In addition to this defensive ability some puffers are also equipped with poisonous spines, and even their bodies contain a very powerful poison which can be fatal to humans.

Another family of poisonous fish is the Ostraciontidae, the box fish. These astonishing fish have a layer of fine bone under their scales so that they are rigid or box-like. In addition they are almost square in cross-section which further helps to create the impression of a box. Some of them feed on algae whilst others feed on small invertebrates. If one is collected with a net and consequently frightened it may release poison, so that if placed in a closed container of sea-water other fish present may be killed. The extent to which it may be able to poison its predators in the sea is uncertain, but presumably such a mechanism could operate only in confined spaces like crevices.

Little can be said in this account about the large number of carnivorous fish which prey particularly on smaller fishes. These predators fall into two groups, those that visit the reefs and those that live all the time on the reefs like the groupers (family Serranidae). Many of the latter are territorial and patrol a fixed 'beat', either by day or by night, in search of prey. Some, like the famous Queensland grouper grow to immense sizes reaching 2 m (6 feet) or more in length. Generally these fish swim slowly towards their victims, and when close enough open their mouths to draw in quantities of water and the prey too. Single groupers of all sizes are to be found in their selected positions around most reefs. A number will take invertebrates for food as well as fish.

Another interesting fish predator is the Scorpion fish or Lion fish *Pterois volitans* (family Scorpaenidae). This fish bears a number of frilly fins and long poisonous spines on its back. It also has long spiny pectoral fins. It is attractively marked and lurks in or near crevices in the reef or among weeds. Small fishes possibly mistake it for a plant and when they swim close to it they are quickly engulfed by the relatively large jaws. In other cases it stalks its prey and then makes a sudden final dash to make the capture.

Tourism and coral harbours

Since the first discoveries of coral islands by Europeans, visitors have come to the islands as traders and more recently as tourists. Some of the voyages ended on the reefs with a crippled ship and a scramble for life, but generally the arrival of ever larger ships has led to the construction of harbours which can have interesting effects on marine life.

The construction of any sort of port or harbour can only take place in appropriate locations with the right conditions of shelter from winds and currents. There must also be access for ships which require depth of draft for safe passage. This means that a coral reef can easily block all approaches to land for vessels of any size unless a suitable channel can be found through it. In some cases, due to natural conditions of substrata or current, corals have not themselves been able to develop equally all round islands, so such an access may be possible on one side at least. In other cases, such as along the coasts of the Red Sea, characteristic breaks in the reefs occur, and these breaks and the associated part of the lagoon form excellent natural harbours, called 'mersas'.

The building of submarine structures like quaysides and pilings affords ample opportunities for investigations into the development of colonies of coral and the growth of other marine organisms. Many concrete structures, quite apart from the many other types of 'drowned' objects such as shipwrecks, provide excellent substrata for coral growths. Initially they become discoloured by microscopic growths like bacteria and algae, and once these have established themselves the surfaces then prove attractive to a whole range of invertebrate larvae which settle from the plankton and search for space. The physical texture of the concrete itself, so long as noxious chemicals have been dissolved away from it, can prove of great importance. Many settling larvae are attracted by rough surfaces which they find easy to adhere to. Because of these characteristics a new range of ecological experimentation has become possible. This seeks to study the problems related to site selection in marine invertebrates like corals by depositing on the sea-bed concrete blocks or stones at selected points. These blocks constitute an artificial reef, and their use has gone some way to showing the sequences of colonization among corals and other invertebrates as well as the preferences of different species for locations with particular directions of illumination and currents.

All but the most simple harbours will have faunas and floras which differ from those of the adjacent 'natural' areas. In some of the larger ports of the tropical world we may encounter quite unusual populations of animals. Because of the shelter afforded inside the harbour and the suitable substrata provided artificially, prolific growths of sessile invertebrates may develop. These in their turn will attract more fish and other animals to feed on them. In many harbours the water, which may still be quite clean chemically, can have very large concentrations of organic material present. This might result from a variety of sources, such as effluent from on-shore waste disposal systems or jetsam from ships. When the conditions are right this material may be quite important in supporting some food

Left: A fisherman in his outrigger canoe makes his way carefully round a reef to avoid damaging his boat on the coral colonies in the shallow water

webs. Accordingly, differing populations of carnivorous and herbivorous invertebrates and fish will be found. Interestingly, some of these may show variations in form and coloration from animals of the same species which occur outside the harbour. Often they grow bigger and sometimes their colour patterns are darker, suggesting that they represent a race of individuals which have been more successful in the tainted darker water. The extent to which their breeding is limited and separated from that of the 'normal' populations beyond the harbour is difficult to estimate.

Not all harbours are the large, well-constructed commercial type, however, particularly among the coral islands. Many coral ports are small and here the living influence of the coral can be felt more strongly. Sometimes it is difficult to keep the navigation passages open. If access to the harbour is through a cleft in the reef, active coral growth may restrict the beam and draft of vessels as corals grow from the bottom and from the sides of the channel. Such a situation was probably responsible for the gradual closing of the channel into the ancient port of Suakin on the Sudanese coast of the Red Sea. Originally this site had been selected because of the natural access through the reef and because of the ease of fortification. In earlier times it was the main seaport for the Sudan, and it survived the comings and goings of many armies and of two empires all of which left their marks on the old city. With the increase in the size of ships the access to the harbour, which apparently continued to be restricted by the growth of coral, was no longer adequate so that shipping had to be transferred to a new port which was built a few miles away at Port Sudan. Suakin is now virtually a dead city with its beautiful Turkish houses falling into decay, inhabited only by cats and bats. Much of the material for building in this coastal region of the Sudan is taken from fossil coral reefs inland, so the buildings themselves have an even older history, reflecting bygone geological eras when sea levels were different from those at present, and when animal communities were probably different too. In the harbour only a few small craft are to be found where once the proud vessels of the Ottoman Empire lay at anchor. Some of the small ships that still come to the harbour rely entirely for their commercial existence on the productivity of the coral reefs and the adjacent seas. Many are used in the small but important fishing industry.

Left: Pearl-oyster fishing with piraguas or dug-outs in the Maupiti lagoon in the Society Islands

Coral reef fisheries vary considerably from those in temperate waters because of the very different types of fish available and the vastly different nature of the sea-bed. Because of the structure of the reefs, which rise up so steeply from the sandy floor, as well as the periodic rocky outcrops of small coral knolls, any form of trawl fishing on the sea-bed is quite out of the question. Expensive trawling equipment would snag on coral outgrowths and be lost. A coral reef, however, is a very productive place in terms of fish population. Very few open water fish live near the sandy bottoms unless they are associated with shallow bays, so generally there would be no commercially valuable fish there. Rather, the coral reef fishermen concentrate their efforts on the outer and inner sides of the reef as well as on the lagoon. To a great extent this fishery is conducted using lines and baited hooks. Only in the lagoon is netting really practicable. Here, in the shallow water, a very important fishery exists for small sardine-like fishes which form the standard bait in many parts of the Indo-west Pacific, and especially in the Red Sea. Shoals of these fishes may be seen swimming in the shallow water to find food, and here they can be easily surrounded with a net. In this manner the fisherman provides himself with bait for his hooks. Some of the more active open water fish, like tunnies and barracuda are taken successfully with spinners trailed behind slow moving boats as they pass along the edge of the reef. Apart from their commercial value, these fish together with some others like sail fish, may form the basis for a moderate sport fishery which can be a great attraction for tourists.

The environmental effects of exploitation for tourism are very difficult to quantify. Tourism frequently has a most profound effect upon the native human populations of coral islands because it often brings them into more or less direct contact with people from very different backgrounds and cultures. The influence of tourism upon a community can be great. Apart from the consumption of natural resources in bringing the tourists to their coral island, a variety of raw materials will be in demand there for supplying new buildings and amenities in places where nothing like this ever existed before. Often men and machines have to be drafted in from great distances to carry out this work. Harbours, roads and airstrips have to be constructed and communication and drainage systems installed. All these activities inevitably have serious effects upon the ecology of the land and to a considerable extent upon that of the sea. In many cases the construction of harbours or the extension of port facilities has had direct effects on the natural communities of the adjacent sea. One example of how this may happen is when dredging and excavation effect the patterns of deposition of natural sediments, thus smothering corals.

Often, however, it is the tourists themselves who have the greatest effect upon the well-being of the marine environment through the influence of various popular activities. The principal dangers here are from pastimes like reef-walking, shelling, spear fishing and scuba diving, especially when the people involved have not been warned of the consequences of unlimited interference with the plants and animals of a reef.

Due to the exhaustive habits of professional and amateur shell collectors who often pass their finds on to the well-established shell markets, many of these beautiful objects are becoming increasingly rare. In many cases the creatures involved take a number of years to reach maturity and thus to reproduce themselves, so that the activities of one collector can have a serious effect on the breeding population in any given area.

The same is true for many coral colonies themselves. Some of the branching genera like *Acropora* grow relatively quickly, as evidence from the recolonization of damaged or artificial reefs has shown. The more massive ones like the brain corals and the genera *Porites* and *Goniastrea* may take much longer to develop, and it is sad to reflect that just a morning spent on a reef by a

Right and below: Two shots of the wreck of the 'Wanderer', a sailing brig, on the Tuamotu barrier reef. Those parts of the vessel below the water are soon encrusted with algae and corals

professional coral collector can do great damage to a population of colonies which have been developing for tens or even hundreds of years.

The fish on a coral reef may also be at risk from a variety of sources. Perhaps the most serious threat is that posed by the spear fisherman who comes to the island armed with powerful spear-guns. Many forward-looking countries throughout the tropical world have passed legislation restricting or banning their use or import. Although under certain circumstances, in open water when the hunter is not using scuba, spear-guns can be regarded as sporting, their use on territorial fish like groupers is far less so. These fish, being territorial, usually return to their favourite crevice or hole in the reef when they are chased, and thus fall easy prey to the determined use of a spear-gun. It would be far more desirable if tourists could be encouraged to abandon spear-guns and bring cameras instead.

The great difficulty in the introduction of protective legislation lies in its enforcement: coral reefs are obviously difficult to police. Governments have also found themselves in something of a dilemma, because strong prohibitive legislation is certainly no encouragement to the tourist to make a visit. On the other hand, if there is no protective legislation there may well be no underwater tourist attraction in a few years time. One way out of this situation has been the policy adopted by some authorities of limiting the activities of collectors and spear-fishermen to certain areas or to certain off-shore reefs. This is perhaps an ideal situation as it allows especially careful protection of outstanding communities.

Another source of conservation problems has arisen from the activities of professional fish collectors who supply the increasingly fashionable marine aquarists' shops of the temperate world. Many of the smaller reef fish are quite easy to catch, and in certain countries collectors have established themselves in quite a large way, often employing a number of native helpers. A proportion of the fish that are exported under such circumstances cannot survive shipment, quarantine, and wholesale and retail sale, so that they are, of course, for ever lost from the breeding population of the reef.

The activities of tourists on the reef and of professional collectors can therefore pose a great threat to the maintenance of a good tourist attraction. However it need not do so. So long as the various productive reef industries are carefully controlled and annual takings limited, a reef should be able to support some exploitation. The position of the tourist is also not immutable. Much can be done to attract, inform and increase

interest by means of a systematic educational approach to conservation. The establishment of a good marine aquarium in some easily accessible position can be a great help. So too can the production of concise field guides which can assist the visitor with the location of certain especially attractive areas, as well as with the identification of some of their more conspicuous inhabitants. These are good starting points. One item which requires very little expenditure is the establishment of a Nature Trail across a conserved reef. This can be designed with either the reef-walker or the skin-diver in mind and would help enormously with the problem.

A lagoon may form the setting for an active shellfish industry. Prawns are one of the more valuable catches, and although they are often caught in open bays, they frequently enter the lagoons and other small inlets such as those associated with commercial salt pans. Salt pans are usually shallow excavations set a little lower than the sea level, and cut off from the sea or lagoon by a bank of earth or sand. When a pan is

Above: These sea-squirts with yellow rings surrounding their exhalent and inhalent siphons, are filter feeders. They are permanently attached to the reef. Also in this photograph are brownish sponges and feathery colonies of hydrozoans

Above: A variety of different growth forms of true coral. In the centre is a dark brown surgeon fish surrounded by yellow and bluish damsel fish

to be filled this bank is breached and the sea flows in bearing with it a variety of organisms including the prawns. These animals can be captured by stretching a net across the breach. When the pan is filled the breach is closed off and the heat of the sun begins the process of evaporation until all that remains is a crust of white salt crystals which can then be cut out of the pan in blocks, crushed and sieved before marketing.

One of the most potentially profitable industries is that of culturing pearl oysters. A great deal of experimental work has been carried out on this topic in order to improve the productivity of the oysters and the quality of the pearls. This work has not, however, decreased the value of the pearls collected by the traditional pearl divers and some coral areas are renowned for their courageous divers and the quality of the pearls they can collect.

The shallow lagoons and natural harbours of many islands provide for another industry based on invertebrates. This is the Beche-de-mer fishery industry. Many species of sea-cucumber inhabit such waters, feeding on the sand itself which they waft into their mouths by means of their feathery oral tentacles. The sand is taken in along with any living or organic material that it contains. This is digested and the 'clean' sand is voided through the cloaca of the animal. The black sea-cucumbers which usually form the basis of the fishery belong to the genus *Holothuria*. After they have been collected they are dried for export and sale, particularly in oriental countries.

One outstanding feature of many islands, as well as mainland shores in coral reef areas, is the accumulation of wrecks of ships which have failed to make a safe passage through the reefs to the shelter of the harbour, and have fallen victim to the sharp edges of the coral rock which rises from the bottom and the sides of navigation channels. An investigation of one of these wrecks makes a fascinating excursion, both in order to explore the decaying ship and to see the organisms which have taken it over, often forming a ghostly ornamentation to the superstructure.

Men of the coral islands

by C. R. Hallpike

The coral islands of the world, scattered in vast numbers throughout the tropical seas, especially in the Pacific, have become famous since the voyages of Captain Cook as earthly paradises. In reality they are often poor, drought-ridden lumps of rock raised only a few metres above the sea, with meagre soil, a scarcity of fresh water, and no wild game, isolated from one another by great expanses of ocean. Yet in this unpromising and demanding environment men have, by ingenuity and determination, and by using every available material to its utmost, created complex societies, become expert fishermen and built long and seaworthy canoes in which they have voyaged over vast expanses of ocean to trade and fight.

After the arrival of the Europeans, the very poverty and isolation of these atolls has been to their advantage, since they have offered little to attract the greediest of the white men. Their small size has often resulted in over-population, however, and for this reason infanticide and emigrations have been common. Since there is usually no wild game worth hunting, the islanders depend on agriculture and fishing. The way of life on all the coral islands of Pacific Oceania is basically similar, in spite of cultural variation between the Polynesian, Melanesian, and Micronesian peoples.

Atoll formation has been discussed elsewhere in this book. As has been said, the typical form of an atoll is a ring of reef around a central lagoon, though the surrounding reef is usually broken into a series of small islets. The Polynesian Tokelau group, for example, consists of three atolls,

Left: A coral islander checks his course with a bamboo navigational aid

totalling 18 square km (7 square miles) of dry land, and the atolls each comprise 63, 93, and 62 islets. Some atolls may be very large, however, and Rangiroa in the Tuamotu Archipelago is made up of 20 islets around a lagoon 72 km (45 miles) long by 24 km (15 miles) wide. Christmas Island, the largest atoll in the Pacific, is 145 km (90 miles) in circumference.

The outer edge of a reef is steep, and since the atoll is on a volcanic elevation the sea bed falls away precipitously beyond the reef to great depths. The central lagoon, however, may be only 18 m (60 feet) deep or less, as on Uvea in the Loyalty Group, where the lagoon has become silted up altogether and provides a fertile swamp. Though coral atolls are basically narrow strips of land, shifts in the zone of coral building activity or the piling up of shingle can both widen the original reefs.

Coral can only rise a little way above the surface of the sea because of wave action, but in some cases it has been raised up to 100m (330 feet) as the land beneath it has been pushed up by pressure in the earth's crust. These elevated islands have low cliffs, separated from the sea by narrow but level strips of land that are highly suitable for agriculture—the island of Maré, for instance, has no fewer than three sets of these cliffs. As we shall see, the considerable depth of limestone on these elevated islands has, in the many cases where sea birds have deposited masses of guano on them, acted as a giant sponge, absorbing the phosphates washed out of the guano by rainwater. This provides millions of tons of rich fertilizer and, in the twentieth century, has attracted mining companies.

The soil on the coral atolls is basically powdered

coral mixed with seaborne sand, which is rendered even less fertile by the fact that in many areas of the Pacific the islands are liable to drought. Even where there is plenty of rain, the limestone allows it to percolate down to sea level, and on the raised coral islands the women have to collect their water from shallow pools found in caves.

Coral islands are therefore utterly different from the high volcanic islands of the Pacific which display peaks and ranges rising sometimes to a thousand metres and more. These mountainous islands are able to tap the rain-bearing clouds of the westerly winds yielding a high annual rainfall. The river valleys of the high islands are clothed in dense vegetation, and the weathering of the volcanic rock produces rich tracts of deep soil in the lowlands.

There are atolls, on the other hand, which will support only grass and plants, though there is usually an abundance of fish and birds. Because of their lack of water they are suitable only as temporary camping sites for native fishermen on their voyages. Coral islands with higher rainfall such as the Marshall and Ellice Groups have a richer soil, supporting coconut, pandanus, and breadfruit trees, besides yams and sweet-potatoes. The islands with the highest rainfall, such as the Gilberts and Tokelau, can sustain the moisture-loving taros and bananas. The population densities of atolls obviously vary in the same way from only 20 or so per square mile to that of the Gilberts which reached 200 per square mile.

Often because of over-population, warfare between different groups on the islands was common, and was exacerbated by European traders who sold iron weapons to the natives so that

Left: Mother-of-pearl from oysters still provides the islanders with an important source of income

some chiefs were able to dominate other atolls. The Gilbertese were especially warlike, and developed a number of ingenious weapons, such as shark teeth embedded in clubs and swords, spears tipped with sting-ray barbs, and coconut-fibre armour. Captives were sometimes kept as slaves or accepted into the community, but cannibalism was common as well.

The coral islands of the Pacific mostly lie in the tropical belts of the easterly trade winds and the westerlies. The easterly trades blow for most of the year, and then after a period of calm the westerly winds bring rain. Wind temperatures remain much the same, however, and since the atolls are so small in relation to the surrounding sea their average year-round temperature varies only a degree or so from 27°C (80°F).

This exceptionally favourable climate means that houses can be of simple construction. In some areas they are raised on piles, while in others they are built on the ground, usually on a rectangular plan, with walls made of matting, and thatching done with leaves.

Because of the great sea distances involved, only a very limited number of plants and animals have been able to penetrate Oceania unassisted by man. The Wallace Line, in particular, running between Bali and Lombol, and Borneo and Celebes, in Indonesia, is the boundary to the east of which the higher mammals did not migrate, so that they failed to reach the Pacific Islands and Australasia. Only birds, bats, and some tree opossums in a few islands in Melanesia can be hunted, though there are also many lizards and insects. Pigs, dogs, cats and rats have all been introduced by man, though some pigs have since reverted to a wild state. The yam, taro, banana, breadfruit, and coconut are all Asiatic in origin,

Below: A Tuamotu islander uses his wooden harpoon, the 'patia', to bring in his catch

Left: Coconuts, carried by ocean currents, can take root and produce full size palm trees on the tiniest specks of land

Below: Pearl-shell divers haul in their basket

the sweet-potato being the only cultivated plant from the American continent.

Taro is probably the earliest domesticated plant in Oceania, and very many local varieties have now developed. It is not found on many of the dryer atolls since it needs a rich and moist soil, and some islanders construct irrigation channels to grow it. Though it cannot be stored, it grows throughout the year, coming to maturity four to six months after planting. The planting itself is a very simple matter of cutting off the stalks and replacing them in the ground. The main edible part of taro is the thick starchy root, but the tender leaves are also eaten as greens.

Yams are more widespread than taro. They are very variable in size, and some specimens can weigh 15 to 20 kg (30 to 40 pounds), and weights of over 45 kg (100 pounds) are not unknown. Unlike taro, yams can be stored for months, which has led many communities to grow more than they require and give away the surplus at ceremonial feasts. Yams, especially in Melanesia, are associated with virility, and women are usually forbidden to have anything to do with their cultivation. A great deal of magic is involved in ensuring their growth, and protecting them against evil influences. First fruit ceremonies are commonly held when the first yams are dug, usually by the men, but the women carry them back to special yam houses in the villages.

Root crops are cooked in earth ovens, pits in the ground in which hot stones are placed and covered with leaves, and many different sauces are prepared using fish and coconut milk to add zest to what would otherwise be a monotonous diet. On some islands the people dig trenches in the coral and fill them with humus in which to grow better crops of tubers.

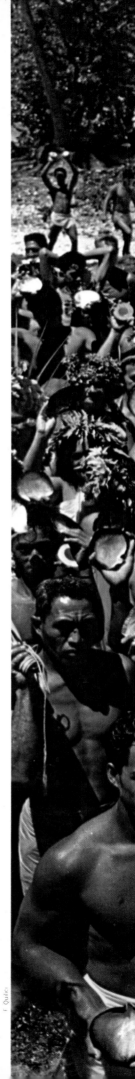

The coconut palm is the most common tree of Oceania, as well as the most useful. Though the seeds are carried for hundreds of miles in ocean currents, they cannot survive for long periods in salt water and the tree must therefore have been greatly spread and propagated by man. It will grow in the light saline sand formed by the disintegration of old reefs, but it needs a well-drained soil and plenty of sunlight. The environment of the coral atoll is therefore ideal for it. The palm matures after eight to ten years, and then lives for about 80 years, bearing about 50 nuts throughout the year.

The mature nut has a thick outer husk which acts as a protective coating around the hard, thin, woody shell. Inside this is the 'meat', an oily white kernel. At the centre of the nut is the milk, a thin, transparent fluid with a tangy and refreshing taste when the nut is green. The milk is

Above: Men cooking their mid-day meal of fish

Right: Islanders celebrate their good fishing by feasting and showing off their best shells

mostly absorbed into the meat as the nut matures, however, so that the nut can be taken green from the tree for its milk, or picked when it is ripe and the meat is harder and more suitable for grating. It is split by removing the outer husk, and impaling the nut on a sharpened stake fixed in the ground. The meat is scraped out with a serrated shell into bowls of water and boiled to a paste which is then either mixed with taro or yams, or used in making sauces mixed with fish.

On some islands, where there are no freshwater pools or springs, the milk of the coconut is the only substitute for water. The oil is pressed from the meat and used as food, as ointment, and as a cosmetic. The fibrous husk provides the raw

material for rope and cord; known as sinnet, the basic cord is made into a string by rolling the fibres across the thigh to twist them, several of the strings then being plaited. Bundles of these sinnets can then be twisted to form thicker ropes.

The shells, when scraped and cleaned out, become cups, flasks, spoons and other utensils. Coconut shell is very hard and therefore highly suitable for decoration, so that many beautiful carved ornaments are made from it throughout Oceania. The trunk of the tree, when it is finally felled or blown down, provides wood for building houses and canoes, and for various utensils and furniture. The leaves are used for thatching houses and weaving mats, either for sitting on, or to act as walls and partitions in houses. In extreme cases of famine and crop failure even the pith inside the trunk can be eaten. In some parts of Micronesia, as a result of Asiatic influence, the flower bud is tapped and the sap, which is full of sugar, drawn off and allowed to ferment, producing an alcoholic drink.

The coconut is essential for survival on many coral atolls; it also enriches the industrial world, and today is one of the islanders' most important sources of income. The dried meat, known as copra, is pressed to provide oil for margarine, cooking and salad oils, soap and nitro-glycerine. It is particularly important in soap making since it has exceptional lathering properties. Copra cake, the residue after the oil has been extracted, is a protein-rich ingredient in animal stock food. The fibrous husks are used for coir matting, and coconut shell charcoal is one of the best absorbents of poisonous gases and odours.

The breadfruit tree is also of great importance though it needs special soil. It matures after five years to yield a starchy fruit, about the size of a

Below left: The men of the atolls are brave and skilful underwater fishermen. They dive and hunt simply by holding their breath, and often swim without masks

Below: Atoll in the Sous-le-Vent Islands of French Polynesia

melon, and continues bearing for 50 years. Since different varieties ripen at different times of year, the islanders can be assured of a constant supply. There are many ways of eating it, but the commonest is to bake the whole fruit in hot embers and scoop out the interior, which is soft and smooth with a taste that has been compared with roast chestnuts, or good bread and potatoes. It can be preserved uncooked in pits, where it ferments with an unpleasant smell, but after baking becomes a pleasant and nutritious food. Breadfruit can also be cooked for storing, by baking it whole in large, stone-lined pits in which fires have previously been lit. The cooked fruits are cut up, dried, and packed away. The excellent storing properties of the breadfruit make it particularly suitable for provisioning canoes for long sea journeys. The fibrous inner bark of young trees can be used to make a type of bark cloth; the wood is used for making utensils and canoes, while a glue and caulking material for canoes can be obtained by cutting the trunk and collecting the milky juice which runs out.

There are other, less important trees that supplement food supplies, such as the pandanus, with its long pulpy fruit, and the canarium almond whose nuts can be smoked and added to yams and taro as flavouring. The areca nut, eaten with burnt coral lime and betel pepper leaves, is both a food and a stimulant. Sugar cane also grows on many atolls, as does the paper-mulberry tree which is a principal source of bark cloth.

The inhabitants of the coral atolls depend far more than those of the high islands upon the sea and are therefore better fishermen. This is because they have no rich hinterland to provide game or vegetable products, and for the most part the

Above: Shell gathering in the lagoon often produces a beautiful specimen that can be sold to tourists

Left: Tuamotu fishermen preserve their fish simply by hanging them out in the sun

soil has no stones which can be used as tools. The great expanses of ocean provide relatively little fish for the islanders who catch what they need closer to home, especially within the lagoons inside the reefs. Even here, however, fishing conditions vary enormously from island to island, and the people have had to expend all their ingenuity in taking full advantage of the sea. Bare hands, rods and lines using hooks carved from either tortoise-shell, mother-of-pearl or clams, bows and arrows, spears and nets are all used. Giant manta rays may be caught with detachable harpoon heads.

In some islands the young of certain species of fish are placed in freshwater ponds where they grow to a great size. Intricate basket-work fish traps are used, and in some places sucker fish are attached to a line and guide the fishermen to larger game. In a variation on this strategy, decoy fish are kept in pools, and when needed are attached to lines to attract others. In the shallow waters of the lagoon large nets attached to poles are used to catch shoals of fish by raising the net at all four corners at once when the shoal crosses it. In deeper waters long nets are suspended vertically from floats, and the men then dive for the fish that are trapped in the meshes. In all these ingenious ways the islanders catch innumerable kinds of fish, as well as shrimp, crabs, lobsters, clams, mussels, eels, octopuses and turtles.

Sea birds are also an important source of food, and on Ocean Island the men capture frigate birds which are then tamed and used to decoy

other birds within range of the owners, who bring them down with a kind of bolas.

Another important food is the palolo, a sea worm, which swarms regularly every year, though at different dates depending on the area. From 15-20 cm (6-8 inches) long, they live in the submarine crevices in the reefs, and once a year the rear half of each worm breaks off and rises to the surface, where it bursts to release its eggs. These segments are scooped up in their millions, and can be dried and smoked over the fire.

In this stoneless environment shell assumed a great importance as the material for tools such as scrapers, adzes, and gouges (though these have long been replaced by steel) and was also traditionally used for currency, ornaments, and trumpets. The most beautiful shells such as giant cowries, cones and gold-lip are still important items of trade.

Oysters yielding mother-of-pearl and pearls are found near reefs and in the lagoons, and the gold-lip shells may be 30 cm (1 foot) in diameter and weigh as much as 5 kg (11 pounds). Native divers, working from small boats and naked except for a stone weight, a knife and a basket descend up to 15 m (50 feet) to tear off the shells from the sea-bed until the air in their lungs is exhausted. The occupation has always been highly dangerous; deaths from sharks, giant eels and sting rays have been common, and lung collapse is a constant danger.

Canoes are therefore essential to the islanders' survival, and have been an integral part of the migration of peoples, and their trade, warfare, and subsistence in Oceania. Canoe building is often a specialized craft, and also a collective enterprise, so the vessels usually belong either to the group or to the chief.

On some islands canoes are or were of dug-out construction, and the only tool used was a small stone or shell adze. By continual burning and chipping, the trunk of a tree—usually coconut or breadfruit—was rounded on the outside, hollowed inside, and pointed at both ends. A simple outrigger is fastened to canoes by lashings through holes made by burning. Occasionally a section of the boat may be decked, and a triangular sail fitted, especially for voyages on the open sea. Sails traditionally consisted of two or three pieces of matting stitched together with a needle made from the wingbone of the flying fox, while a large stone served as the anchor.

On other islands canoes are of plank construction, which requires the splitting and adzing

Left: The coconut palm in two stages of development. A young tree is shown on the left, a mature tree on the right

of the keel and at least three planks on either side. Holes are bored along the edges of the planks (with flint or shell augers before metal tools became available) and the hull is built with temporary lashings. Cross-struts are also fitted, and the boat is left until the timbers have bent themselves to shape, after which it is dismantled for the final adzing and polishing with coral rubbing stones and fish skin stretched over blocks of wood. The planks are then given their final lashings of sinnet cord or creeper, and the bow and stern made higher with more planks. Some canoes were elaborately decorated with shell ornaments and painted designs, and with carvings of birds, fish, and geometrical designs, especially on the prows. Plank canoes, being wider than dug-outs, are more stable and outriggers are correspondingly less necessary. Nevertheless they are usually fitted, except in the Solomon Islands. On sailing voyages the freeboard was increased, and canoes were also propelled by paddles. The largest double-canoes could hold over 100 adults with provisions.

Though the islands produced the bare necessities for survival, many manufactured some speciality, such as shell ornaments, dyes, or finely woven mats, and trading voyages were made over many hundreds of miles to obtain pottery, jade, tortoise-shell and other precious objects, often of little practical value. It was possible for whole communities, with their pigs, fowls, and food plants, to set out together in fleets to search for new land if their homes became overcrowded, or if famine or drought made them uninhabitable. Revictualling at temporary resting places on small atolls, the fleets could spread out over a wide front so as not to miss some distant speck of land that might provide a permanent refuge.

Below: On the island of Trinidad the coconut palms extend almost to the beach itself

Above: A ripe coconut, still in its green waterproof skin

Adventure, prestige, and exploration were also important motives for these voyages, especially among the Polynesians. According to tradition, for example, canoes from the Society Islands had travelled to the furthest limits of Polynesia, to Hawaii in the north, and New Zealand in the south. These great voyages were made possible by the seasonal reversal of the winds and the use of different ocean currents, but they also depended on an expert knowledge of the stars. The Marshallese, the most daring and skilful navigators in Micronesia, departed on voyages lasting for many months, taking with them charts made of strips of wood, held together with fibres, showing the islands and prevailing winds.

Coral atolls thus imposed certain limitations on social organization and daily life, but gave other freedoms impossible to those who lived in the hinterlands of larger islands. Moreover, the

Right: The breadfruit tree will not grow in ordinary coral soil. On many atolls the islanders have prepared special soil pits in which to cultivate them

Above: Heads crowned with flowers, the inhabitants of Bora-Bora in the Society Islands drag their nets full of fish to the shore

Left: A man prepares grated coconut for cooking

cultural traditions of the Micronesians, Melanesians, and the Polynesians have been very important in affecting the different uses to which the various peoples have put the same basic coral island habitat.

The Polynesians, whom many anthropologists consider to be basically Caucasian with a southern Mongoloid admixture, are tall, well-developed, light brown in colour for the most part, with wavy dark brown or black hair, high bridged noses, and very little beard. In the course of their migrations from South-East Asia they have covered the whole extent of the eastern Pacific, from Hawaii in the north, to New Zealand in the south-west, and Easter Island in the south-east.

Despite the vast distances which separate them, Polynesian societies and languages have remained surprisingly similar to one another. Hereditary chiefs led their kinsmen in warfare, feasting, and other social activities, and also in the worship of the group's ancestral deities. The spiritual power of the chiefs and of other sacred things was known as *mana*. Each group had an open air temple, usually consisting of a terrace with stone pillars, where ancestors and other deities were worshipped.

In some areas kinship remained the basis of social organization, in others rank and class distinction surpassed it in importance, and in others loyalty was due to place of residence. Whatever the basis of group organization, corporate control of the individual was strong. Communal ownership of property, especially land, was the norm, and there was a well-developed sense of rank. Nevertheless, it was possible for men of ability to rise in the social scale by their own efforts. Women had high

status, and old age was revered. Unlike the Melanesians, the Polynesians were not greatly interested in accumulating wealth simply for prestige, and their creative energies went into constructing elaborate systems of etiquette, memorizing long genealogies of their ancestors, celebrating complex religious rituals, and elaborating their mythology. Each society took a particular cultural theme and developed it, such as the *kava* ceremony on Samoa, religious ritual on Tahiti, and the cult of birds and the carving of stone statues on Easter Island.

The Micronesians are smaller and lighter in build than the Polynesians, with more delicate features and fairly prominent cheek bones and chins. They have scanty beards, and their hair is straight and black, lank and not wavy like that of the Polynesians, a Mongoloid characteristic that is also found in the oblique eyes of some populations. The Micronesians of the Gilberts have a Melanesian strain, which is also found in the people of the Carolines and the Marshall Groups. There are about eight different culture areas in Micronesia, making them less socially homogeneous than the Polynesians even though they mostly inhabit low coral atolls, but generalizations can be made about Micronesians to some extent.

Micronesian societies are usually matrilineal, that is, descent is reckoned through the mother, not the father, and groups are more clearly and rigidly defined than in Polynesia or Melanesia. Status is more dependent on birth and less on individual ability, and relations between groups tend to be fixed, though there is some outlet for ambitious men to gain prominence by acquiring specialized skills or by competitive feasting.

We therefore find complex systems of social stratification, sometimes of nobles and commoners, or nobles, lesser nobles, and commoners, or, as on Yap very much more complex systems in which nine social classes are grouped into seven ranks, three in the upper class and four in the lower, the upper and lower classes not intermarrying. Chiefs tend to be autocrats, and to marry only among their own class, displaying more rigidity in this respect than the Polynesian chiefs. Settlement patterns are very varied, with some societies living in dispersed family homesteads while others are congregated in permanent villages of some size. Monogamy is the usual form of marriage, and even when polygamy occurs the first wife has great authority in the household. While fidelity is expected in spouses,

Left: Eels are highly esteemed by Pacific islanders, who catch them with harpoons

Above: The Pacific islands are never without the cries of seabirds. Shown here are three blue-footed boobies

complete sexual freedom is allowed before marriage, and in some cases girls live in the men's houses where they serve a kind of sexual apprenticeship. The Micronesians are as skilled in navigation as the Polynesians, and developed the same wide network of trade; they are also the only people in Oceania who traditionally practised pottery and weaving with looms.

The Melanesians are Oceanic negroids, dark brown in colour (the Buka of the Solomon Islands being among the blackest people in the world); they have heavy brow ridges and their hair is frizzy, but their lips are not thick or protruding, and their noses are well-defined without the flattened nostrils of the African negro. They live on the archipelagoes near New Guinea, such as New Caledonia, the Solomons, and the Bismarcks, inhabiting mainly high volcanic islands and not the coral atolls, and there is thus a far greater degree of variability in their habitats and social organization than in Micronesia. The social variations have been intensified by the migrations of many different groups within Melanesia, so that few generalizations about their society and culture are possible. It can be said, however, that they tend to rely on root crops rather than on tree products, that wealth is sought as a means of acquiring prestige through lavish feasts and exchanges of gifts, and that individuals win status by displaying powers of leadership and political skill rather than by birth. Women have a noticeably lower status than in the rest of Oceania. In all these respects Melanesian societies also resemble many of the societies of New Guinea.

After the early explorers, the first Europeans with whom the islanders came into contact were missionaries and traders, especially during the eighteenth century. Apart from some sixteenth-century Spanish missions in parts of Micronesia, the first missionaries in Oceania were British Protestants of the London Missionary Society, and they and other missionary societies soon made a great impact on native society, banning nudity and forcing the women to conceal their bodies beneath long 'Mother Hubbard' dresses, suppressing cannibalism and infanticide, translating the Bible and spreading literacy.

They were soon opposed by Catholic missions, with the support of the French government and its navy, bent on supplanting British influence in the Pacific. The rivalry between Catholics and Protestants was eagerly taken up by their native converts, who were still burning each other's churches in the 1930s! But despite one and a half centuries of evangelism, even by 1939 only about

Right: The manta ray, despite its size and unusual appearance, can be approached with ease and will never attack man

Below: From time to time sharks pose a threat to divers, but attack less often than is popularly imagined

750,000 of the 2,000,000 population of Oceania had been converted, and some of these practised rather bizarre forms of Christianity. The missions were responsible for destroying much of native culture, but they should be credited with opposing many forms of exploitation, against which they were often the natives' only champions. Well into the latter half of the last century they opposed any white settlement, but finally came to realize that it was in the islanders' own interests to have the protection settlement would bring against the worst sorts of Europeans.

The first traders in the early nineteenth century were the whalers of the American, British, and French fleets who used the islands as bases to obtain fresh fruit and vegetables, and the native women as entertainment for their men. The skippers bartered knives, calico, rum and muskets for these favours, but the social impact of several hundred crewmen from the whaling fleets, coming ashore for the first time after many months on to the small coral islands can be imagined.

The increase in shipping in the nineteenth century, the beginning of extensive trade between China and America and the economic growth of Australia and New Zealand opened up the whole of Oceania to traders in mother-of-pearl and pearls, the Beche-de-mer used in Chinese soups, and sandalwood for incense. Diving for shell and pearls became the economic mainstay of many atoll dwellers, and many of the shell beds were completely cleaned out, though since the last war there has been some recovery.

Atolls are not suitable for growing large amounts of sugar, but the enormous increase in sugar output, especially from the great estates in Hawaii and Queensland in Australia, meant that

Right: After a long struggle a giant grouper is finally brought ashore

Below: Fishermen prepare to hoist sail on their canoes and spend another day at sea

Left: The villagers display their best coral for sale to tourists and Chinese merchants

Right: Boys returning with baskets of coconuts which they have gathered from a neighbouring atoll

Left: The preparation of a porcupine fish for sale requires a great deal of patience and skill

native labour had to be found to work the plantations. The hunt for labour led to 'blackbirding', one of the worst abuses ever perpetrated on the Pacific islanders. They were lured on board ship by promises of earning great wealth on the white man's plantations and being returned home as men of substance after a few years. Many were kidnapped never to see their homes again, while others were paid off in rum and muskets with which they caused havoc when they were finally brought back. In the east the Peruvians seized whole island populations and sold them as slaves to work in the Peruvian guano mines and plantations, where most of them died. These activities were not finally abolished until the end of the nineteenth century.

The growing needs of developed agriculture throughout the world produced a heavy demand for phosphate fertilizer, and the atolls on which there were deposits of guano were soon scoured. The phosphate impregnated rocks of the raised coral islands such as Ocean, Nauru, Makatéa, and Angaur, however, encouraged large mining operations, and so rich were the deposits that some are still being worked to this day. The inhabitants of these phosphate islands soon became dependent on the mining companies and their old social organization rapidly disappeared. But they were fortunate in comparison with some islanders, since they were paid royalties by the mining companies, and the overcrowded people of Ocean Island, for example, used this money to buy an uninhabited island in the Fiji group and settle it.

It was the growing world demand for copra, however, that made the greatest impact on the way of life in the coral islands. Unlike sugar or phosphate mining, copra can be produced by economic units of any size, and needs only to be

Right: 'Our land is the sea' is a phrase from the islands. Most of the islanders' needs are provided by the sea

Above: An islander checks the stock in a fish-farm

Left: Fish reared in a simple fish-farm are harpooned and tossed into nets

dried and bagged to be ready for shipping. While the coconut plantations attracted large numbers of independent white settlers to the islands, and in a few places huge plantations were run by professional managers with large European staffs, the natives themselves could enter this market as producers and earn the money to buy all the western goods which a century and more of contact with Europeans had given them a taste for. By 1939 Oceania produced an eighth of the world supply. Though the European plantations were responsible for the loss of more native land than any other factor, copra has been the most important source of economic independence for the islanders.

Events during World War II caused great social dislocation in the islands, especially in Micronesia. Here the Japanese used the atolls as sources of fish, copra, sugar and phosphates, and forced the islanders to work as coolies in the construction of a vast network of bases, though the inhabitants of the remoter atolls were mostly left alone. Heavy fighting occurred in this area, and at the end of the war many Micronesians found their boats all destroyed, their domestic animals killed, the fish of the lagoons and reefs wiped out by naval bombardment, and their gardens and groves threatened by insect pests introduced by the Japanese.

The Allied powers did their best after the war to repatriate the labourers deported for forced labour, to give medical and economic aid, and to restore such representative government as had existed. After some years the islands began to recover, especially since copra prices were maintained on world markets.

There were great population increases, however, and some peoples, especially from the Gilberts, had to be resettled. As shipping increased again there was more migration between the islands and men travelled further afield in search of work. The increasing number of tourists and the islanders' growing awareness of the outside world sharpened political discontent.

There is now a powerful tide towards independence for most of the islands, except for the most remote and poverty stricken. Economically, the future will depend on cash crops under the islanders' own control, and they can look forward to growing revenues from tourism, wherever atolls are not too barren and remote to attract visitors.

Protecting the coral heritage

It is becoming more and more widely recognized that man's continual attempts to improve his standard of living rarely have beneficial effects on the environments surrounding him. In many Western and some Eastern countries of the world, increasing industrial development has resulted in devastation of adjacent areas of land in the search for more and more natural resources or 'raw materials'. Shifts of population towards places where work may be obtained have led to the spread of conurbations. In the face of this onslaught, natural communities of plants and animals retreat farther and farther from the places in which they once used to thrive. Sadly this is a situation which has already repeated itself many times on the continents of Europe and America.

The shift towards a more industrialized society and the concentration of populations of working people near their places of work have naturally raised considerable problems. Some of the most outstanding ones relate to the ever increasing shortage of natural resources, while others concern issues like waste disposal. It is hardly surprising that mankind has in recent years looked more and more to the sea for solutions to these problems. The sea does offer an area for activities where many of the less environmentally acceptable of them will be 'out of sight'. In a sense then, exploitation of the sea for raw materials and waste disposal is quite attractive to authorities who are faced with planning and conservation issues on the land. The attractiveness grows as the problems of expanding populations and shortages of space increase.

Left: A striking picture of Bora-Bora fishermen stretching their nets to dry in the twilight

Conservation is an issue which has been brought increasingly before the public in recent years. Anyone who has experienced the beauty and splendour of life on a coral reef at first hand would find it difficult not to agree that such an environment should be preserved. Sadly, today these fascinating and complex communities of plants and animals are becoming increasingly at risk from the pressures of man's expanding activities. This appears to be the case at a time when several authorities believe coral reefs to be in a state of natural decline. It will therefore probably become more and more difficult for the present generations to keep in trust for those of the future the heritage of life on coral reefs. Yet, if preserved, this heritage could be regarded in itself as supporting properly administered industries of fishing and tourism, quite apart from its educational value. It might also prove to be a source of raw materials as yet undiscovered, such as pharmacologically active chemicals.

What is crucially important today is that the developing countries of Africa and Asia, who are so fortunate in their natural endowments, should not squander or pollute them as their European and American counterparts have tended to do in the last two centuries as a result of bad planning and bad industrial administration. This need for caution is all the more important when one considers the increasing speed with which technology is spreading. It is easy to understand how impatient developing countries must feel when they compare their standard of living with that of their Western counterparts and it is only natural that their attentions should turn to their most available assets and the most economical means for their exploitation.

F. Quilici

For the men of coral islands, the most apparant natural resources relate in almost every case to their coasts and their coral reefs. In the past these have produced food and allowed minor commercial acivities such as the collection of shells or pearls. A coral reef can be exploited in a number of ways, and there is no reason why it should not support a certain amount of local industry. The great danger comes when the level of exploitation exceeds the rate of natural replacement. Once this situation is reached irreversible changes may begin to take place, and the splendours of the natural reef population may decay, at first slowly and then more rapidly as the complex inter-relationships existing between the members of the community are increasingly broken up.

The fishing industry, as it has developed around many coral islands, is somewhat self limiting. This is because the demand for fish itself is fairly even, and the most practical ways of catching the fish are not those which permit vast hauls to be made. As has been explained, trawls and other systems of netting are not used easily over coral bottoms, and most operations with nets have to be limited to sandy bays and lagoons. Most fish are taken on the reefs from small boats with lines and hooks. Consequently there is not the threat to the fish stock that is posed in some temperate waters where large fleets of trawlers or drifters can catch tons of fish in a few hours. Further, although the reefs are fairly productive, the types of fish suitable for food are somewhat restricted, and some of the best eating species are slow growing.

To some extent similar problems control other reef industries. The increase in the use of plastic for shirt buttons has virtually extinguished the trade in the shells of sea-snails. The demand for Beche-de-mer (the black sea-cucumbers of the genus *Holothuria*) as food has actually decreased in some parts of the world where it was formerly considerable. Artificial culture of shellfish is to a large degree a new reef industry, and so long as it does not seriously interfere with the balance of the ecosystem it should not pose a threat.

With all these forms of industry it should be possible to organize a system of monitoring in order to check their effects on the environment at regular intervals and whether or not they have decreased or increased in intensity.

In those positions where some damage has already occurred through pollution or naturally occurring disasters, much can be done to make matters better. Recent research has shown that it is possible to 'graft' coral colonies from one place onto another ecologically similar position on

Above: A flock of sea-birds at dusk circling over their nesting site on an atoll

Above right: The channel between two small islands surrounded by wide coral reefs, off Raiatea Island

Right: Lagoon and reef on Apataki Atoll in the Tuamotu Archipelago

damaged reefs. Clearly it is easier to relocate small colonies than to move large ones, but in a very few instances really large colonies of corals such as *Porites* have been shifted considerable distances to repair reef damage. An instance of such a move is that on Green Island, off the coast of Queensland Australia, where corals were grouped around the underwater observatory.

So far only those threats to the well-being of the reef community which originate from the land or the reef area itself have been discussed. But other hazards originate from the sea. Foremost among these is the issue of pollution of seawater by oil and other substances carried in ships. Recent accidents at sea have made the point that even the largest supertankers are vulnerable to shipwreck, so that the risk of very serious oil pollutions is greater than ever due to the vast cargoes that are now being carried. Experience in temperate waters has shown that the threat from oil pollution is bad enough, but that the damage to marine organisms resulting from the use of detergents, applied afterwards to disperse the oil, is even greater. In Europe and America detergents have been applied to oil-polluted beaches with the principal purpose of cleaning up the shore so that it is possible for holidaymakers to use it. From the point of view of the animals who may be poisoned by detergents this can be disastrous. Oil itself, although harmful, is often not so toxic. Some progress has been made with containing oil-slicks by means of booms and sucking the oil away, but this can only be done when the weather is right and where there is the necessary equipment. In the cases of nearly all coral reefs, the distances involved and the lack of facilities make such a programme virtually impossible so that such areas must be regarded as being at high risk.

Apart from oil spills there are other pollution problems in the sea. One example is the accumulation of pesticides. In many tropical regions, because of the vastness of the areas under cultivation, pesticides have to be sprayed from the air. Inevitably, they find their way to the seas, either directly through the rivers or indirectly by way of the processes of food distribution and waste disposal. Once in the sea they gradually become incorporated into the tissues of marine animals, particularly among carnivores since they consume and accumulate the chemicals gathered by other organisms.

Those coral reef areas which lie on or at the edge of the continental shelves may suffer from yet another source of risk. They are effectively situated over shallow bottoms which sometimes

Left: Thickly growing colonies of coral and algae at the junction of the reef and the lagoon on Moorea Island

Above: Branching corals provide a habitat for a variety of fish on the seaward slope of the reef

contain, as is the case in the Great Barrier Reef Province of Australia, a wealth of minerals. Where this is the case these natural deposits lie within man's reach as a result of new techniques of submarine drilling and mining. The utilization of natural resources to the maximum benefit of the economy is a political feature of almost all developing and so-called developed countries. Sometimes the total exploitation of these resources is advocated regardless of the environmental consequences, now and in the future. In situations like this there is generally a major confrontation between the commercial lobby on the one hand and the conservationists on the other. This may take place in an atmosphere which is highly charged and polarized towards the two extremes with little or no common ground. Such a situation often leads to an unsatisfactory result, with outright victory for one side or the other. It appears to be an inescapable fact of life that some tapping of natural resources must be permitted if man is to maintain his standard of living and to develop it. At the same time there is a more hopeful future for the natural environment if these developments can be planned carefully and well in the light of our increasing understanding of natural phenomena, and in such a way that they occur only where strictly necessary and where they will do the least harm.

The coral reefs of the world provide a rare opportunity to appreciate the sophisticated level of interactions between members of the plant and animal communities. The result of scientific researches has greatly increased our knowledge of these relationships, but a great amount remains poorly understood or not understood at all. It is of immense importance that the future of these communities is safeguarded so far as is possible and that the risks from over exploitation, pollution and physical damage is kept to a minimum. The need for increasing watchfulness over the heritage of coral reefs is accentuated by the beliefs of some prominant marine biologists that coral reefs in the world are naturally tending to die out, and that the areas where coral flourishes at present are gradually diminishing. If these claims are true then they add more urgency to the task facing the conservationists. Perhaps the most hopeful sign is the greatly increased awareness of environmental problems which is sweeping the globe at the moment. It can be hoped that such an awareness will generate interest in programmes to educate others in the same way, and to increase the sense of responsibility among governments and other authorities who have charge over such areas.

Bibliography

Atoll Research Bulletins Vol. 1, 1951, Pacific Science Board, National Academy of Science, Washington D.C.
Bennett, I., *The Great Barrier Reef*, Lansdowne Press, Melbourne 1971; Scribner, New York 1974; Warne, London 1973.
Dana, J. D., *Corals and Coral Islands*, Dodd Mead and Co., New York 1872.
Darwin, C. (1842), *The Structure and Distribution of Coral Reefs*, Smith, Elder & Co., London, and Peter Smith, New York 1946–9.
Gardiner, J. S., *Coral Reefs and Atolls*, Macmillan, London and New York 1931.
Jones, O. A. and Endean, R. (Eds.), *The Biology and Geology of Coral Reefs*, Academic Press, London and New York 1973.
Marshall, T. C., *Fishes of the Great Barrier Reef and coastal waters of Queensland*, Angus and Robertson, Sydney 1964.
Scientific Reports of the Great Barrier Reef Expedition 1928–29, British Museum, London.
Stoddart, D. R. and Yonge, C. M. *Regional Variation in Indian Ocean Coral Reefs*, Academy Press, London and New York 1971.
Yonge, C. M., *A Year on the Great Barrier Reef*, Putnam, London and New York 1930.

Index

Page references to illustrations are printed in *italic* type.

Acanthaster (genus) 21, 22, 23, 24, 25, 26, 27, 29, 67, 68, 70
Acanthaster planci 14, 15, *16*, *18*, 21-27, 29, 58, 61, 67-70
Acanthuridae (family) 75
Acropora (genus) 18, *18*, 25, 27, 29, *36*, *37*, 41, *47*, 48, *51*, 61, 62, 67, 68, *69*, 70, 77, 88
Actiniaria (order) 32
Ahermatypic corals *32*, 37
Alcyonacea (order) 32
Alcyonarians 32, *32*, 33
Amphiprion (genus) *81*
Amphiprion bicinctus 68
Angel fish 78
Anthias (genus) 69
Anthozoa (class) 31, 32
Antipatharia (order) 32
Apataki Atoll *125*
Apogon (genus) 76
Arabia 55
Aspidonotus taeniatus 80; *see also* Sabre-toothed blenny
Atlantic 21, 31, 39, 42, 67
Atoll 7, 10, *11*, 14, 43, 44, 52, 93, 94, 95, 101, *107*, 111, 112, 114, 118, 121, *124*
Australia 22, 24, 42, 44, 46, 55, 114, 126, 127

Bali 95
Banana 94, 95
Barracuda 87
Barrier reefs 43, 44
Beach rock 46, 49
Beagle, H.M.S. 7
Beche-de-mer 91, 114, 124
Betel pepper 101
Bismarck Islands 112
Black corals 32
Blenny 80
Blue-footed boobies *112*
Bolbometropon muricatus 78
Bora-Bora *109*, *122*
Borneo 95
Box fish 83
Brain corals 15, *44*, *45*, 65, 88
Breadfruit 94, 95, 100, 101, *107*
Brittlestars 51, 62, 68
Buka 112
Butterfly fish *74*, 78

Canarium almond 101
Canthigasteridae (family) 83
Caribbean 21, 41
Caroline Islands 25, 111
Cassis (genus) 67
Chaetodon (genus) 74
Chaetodontidae (family) 78
Challenger H.M.S. 10
Charonia (genus) 29
Christmas Island 93
Chromis (genus) *36*, 61, 76, *69*
Clown fish *63*, 68, *75*, 81
Cnideria 31
Cniderians 31, 32
Coconut 94, 95, *96*, 100, *104/5*, *106*, *108*, *117*, 121
Cook, Captain 21, 93
Coralline algae 7, 17, 53

Corallium rubrum 32, *34*, *39*; *see also* Red coral
Coral Reefs 43
Crown-of-thorns 14, 15, *16*, *18*, *19*, 21, 22, 23, 25, 26, 29, 58, 61, 62, 68
Crustaceans 68, 70, 76

Dahlak Archipelago 55
Daly Glacial-Control Theory 10
Damsel fish *36*, 61, 68, 76, *76*, *77*, *91*
Dana, J. D. 10
Darwin, Charles 7, 10, 43, 44, 46
Dascellus aruanus 61; *see also* 'Humbug' fish
Dascellus trimaculatus 68; *see also* Domino damsel fish
Diadema (genus) 67
Diadema antillarum 67
Diadema setosum 29, 67, 70
Diodon hystrix 76
Diodontidae (family) 83
Domino damsel fish *50*, 68, *79*
Donkey fish 29, 77

Easter Island 109, 111
Echinostrephus molaris 65
Echinothrix diadema 71
Eel 103, 105, 110
Elkshorn coral 18
Ellis and Solander 21
Ellia group 94
Endeavour H.M.S. 21
Erskin Island 23
Eunicella verrucosa 35

Fan worm *62*
Favia (genus) 25, 62
Feather-stars *60*, 62, 70
Felidu Atoll 28
Fiji 118
Fire-coral *15*, 31
Flying fish 56
Flying fox 105
Forcipiger flavissimus 79
Forcipiger longirostris 79
Fringing reefs *24*, 43
Fungia (genus) *49*, *51*
Fungia actiniformis 43

Gaimard 7
Galaxea (genus) 37
Giant grouper 115
Giant manta ray 76, 77, 103
Gilbert Islands 94, 111, 121
Goat fish 78
Goniastrea (genus) 26, 62, 88
Goniopora (genus) *17*, *42*
Gorgonacea (order) 32
Great Barrier Reef *8*, *9*, *19*, 22, 24, 33, 44, 46, *71*, 127
Great Barrier Reef Expedition 22
Great Triton *18*, 19
Green Island 126
Groupers 83
Gulf of Aden 55
Gulls *53*

Hawaii 107, 109, 114
Helmet-shell 67

Hermatypic corals 37
Heterocentrotus mammillatus 70
Hexacorallia (group) 32
Holacanthus arcuatus 78
Holocentrus rubus 82
Holothuria (genus) 91, 124
'Humbug' fish 61, 77
Hydra (genus) 31
Hydroids 62
Hydrozoa (class) 31, *43*

Indian Ocean 21, *24*, 29, 43, 46
Indonesia *26*, 95
Indo-west Pacific 21, 22, 25, 32, 42, 68, 87

Japen Island *6*
Jellyfishes 31, *67*, 70

Labridae (family) 79
Lagoon 7, 11, *11*, *13*, *24*, 26, 43, 44, 51, 52, 55, 56, 56, *57*, *59*, 68, 85, 87, 90, 91, 93, 103, *103*, 105, 121, *125*, 126
Lion fish *82*, 83
Linnaeus 21
Lithophyllum (genus) 53
Lithothamnion (genus) 17, 53
Lombol 95
London Missionary Society 112
Long Key Underwater Reserve 20
Loyalty group 93

Madreporaria (order) 32
Madrepores 7, 31, 35, 38
Makatéa Island 118
Maldive Islands 28
Mangrove *10*
Manta ray *113*
Maré 93
Marshall group 94, 111
Marshallese 107
Maupiti lagoon *86*
Mediterranean 31, 32, 39
Melanesia 95, 97, 111, 112
Melanesian 93, 109, 111, 112
Micronesia 100, 107, 111, 112, 121
Micronesian 93, 109, 111, 112
Millepora (genus) *15*, 31
Molluscs *18*, 51
Montipora (genus) 42, 53, 65
Moorea 59, *79*, *126*
Moray eels 83
Mullidae (family) 78
Murray, Sir John 10
Mycedium (genus) *68*

Nauru Island 118
New Britain *10*
New Caledonia 112
New Guinea *6*, 44, 112
New Zealand 107, 109, 114
Nullipores 52

Obelia 31
Ocean Island 103, 118
Oceania 93, 95, 97, 98, 100, 105, 112, 114, 121
Octocorallia (group) 31, 32
Organ-pipe coral 32
Ostraciontidae (family) 83

Ottoman Empire 87
Oysters 91, 94, 105

Pacific Ocean 21, 42, 43, 46, 93, 94, 95, 109, 112, 118
Pacific Islands 95, *112*
Palau Islands *36*, *46*
Palolo 105
Pandanus 94
Paper-mulberry tree 101
Parrot fish 75, 77
Pelagic fishes 56
Periclimenes soror 61
Peyssonnel 31
Phytoplankton 39, 56
Pistol shrimp 68
Plankton 29, 38, 56, 62, 65, 70, 76, 77, 85
Pliocene epoch 53
Pocillopora (genus) 53, 65, 68
Polynesia 107, 111
Polynesians *11*, 93, 107, 109, 111, 112
Polyp 7, 11, 14, 15, 16, *17*, 18, *18*, 19, 21, 23, 27, *28*, 29, 31, 32, *32*, 33, *33*, 34, *34*, 35, *35*, 37, 38, 39, *42*, *43*, 56, 61, 65, 67
Pomacentridae (family) 76
Porcupine fish *76*, *118*
Porites (genus) 62, 88, *126*
Porites lutea 51
Porolithon (genus) 53
Pseudobalistes flavimarginatus 67
Pterois radians 82
Pterois volitans 83
Puffer fish 75, 83

Quay 7
Queensland 22, 46, 55, 83, 114, 126

Rabbit fish *74*
Raiatea Island *13*
Rangiroa 93
Rays 77, *80*, 105
Red coral 31
Red Sea 19, 26, 29, 32, 43, 55, 56, *64*, *65*, *66*, 68, *69*, 71, 74, 78, *82*, 85, 87

Sail fish 87
Sabre-toothed blenny 86
Samoa 111
San Blas 25, *40*, 76
Sand cay 14, *22*, *23*, 46
Sarcophyton (genus) *33*, *47*
Scarridae (family) 77
Scorpaenidae (family) 68
Scorpion fish *82*, 83
Scyphozoa (class) 31
Sea-anemones 7, 31, 32, 35, *50*, *63*, 68, *75*
Sea-cucumber 91, 124
Sea-fans *17*, *30*, 32, *34*, *35*, *40*, *46*, *52*, 71
Sea-fir 31
Sea-grasses 51
Sea-mats 15, 62
Sea-scorpion *78*
Sea-slugs 21, *60*
Sea-snail 65, 124

Sea-squirt 15, *48*, *90*
Sea-urchins 16, 21, 29, 62, 65, *67*, *67*, 70, 71
Seaweeds 17, 19
Sea-whips 32
Serranidae (family) 83
Shark *70*, *80*, 105, *113*
Shrimps 61, 68, 103
Slate-pencil urchins *67*, 70, 71
Society Islands *13*, *59*, *79*, *86*, 107, *109*
Soft corals 32, *32*, *47*, 62, *64*, 67, 79
Solomon Islands 112
Sous-le-Vent Islands *101*
Sponges *39*, 62, *64*, *65*, 78, 90, 93
Squirrel fish 71
Stagshorn coral 18, 53
Starfish 16, 21, 23, 25, 26, 27, 29, *55*, 58, 61, 62, 67, 68, 69, 70
Stoicactis (genus) 68
Stone fish 68, *78*
Stony coral 31, 32
Suakin 87
Suakin Archipelago 55
Submerged Bank Theory 10
Sudan 87
Sugar cane 101
Sulu Archipelago *12*
Surgeon fishes *40*, 75, 76, *91*
Sweet-potato 94, 97

Tahiti 111
Takpoto Island *21*
Taro 94, 95, 97, 98, 101
Tetraodontidae (family) 83
Thalissa (genus) 51
Tokelau Islands 93, 94
Tridacna (genus) 61
Trigger fish 67
Trinidad *106*
Tropics 7
True corals 7, *17*, 31, *39*, *42*, *48*, *62*, *68*, *91*
Trumpet triton *18*, 19
Tuamoto 95, *102*
Tuamotu Archipelago *21*, 93, *125*
Tube-worms 62, 65
Tubipora musica 32; *see also* Organ-pipe coral
Tunnies 87

Uvea 93

Vermetus (genus) 65
Visayan Sea 27
von Chamisso 7

Wallace Line 95
'Wanderer' *88-9*
Whale shark 77
Whelk *18*
Wrasses 73, 79, 80

Yams 94, 95, 97, 98, 101
Yap 25, 111

Zooplankton 39, 56, 70
Zooxanthellae 37